Cellular Automata Transforms

Theory and Applications in Multimedia Compression, Encryption, and Modeling

MULTIMEDIA SYSTEMS AND APPLICATIONS SERIES

Consulting Editor

Borko Furht
Florida Atlantic University

Recently Published Titles:

COMPUTED SYNCHRONIZATION FOR MULTIMEDIA APPLICATIONS, by Charles B. Owen and Fillia Makedon
ISBN: 0-7923-8565-9

STILL IMAGE COMPRESSION ON PARALLEL COMPUTER ARCHITECTURES, by Savitri Bevinakoppa
ISBN: 0-7923-8322-2

INTERACTIVE VIDEO-ON-DEMAND SYSTEMS: *Resource Management and Scheduling Strategies*, by T. P. Jimmy To and Babak Hamidzadeh
ISBN: 0-7923-8320-6

MULTIMEDIA TECHNOLOGIES AND APPLICATIONS FOR THE 21st CENTURY: *Visions of World Experts*, by Borko Furht
ISBN: 0-7923-8074-6

INTELLIGENT IMAGE DATABASES: *Towards Advanced Image Retrieval,* by Yihong Gong
ISBN: 0-7923-8015-0

BUFFERING TECHNIQUES FOR DELIVERY OF COMPRESSED VIDEO IN VIDEO-ON-DEMAND SYSTEMS, by Wu-chi Feng
ISBN: 0-7923-9998-6

HUMAN FACE RECOGNITION USING THIRD-ORDER SYNTHETIC NEURAL NETWORKS, by Okechukwu A. Uwechue, and Abhijit S. Pandya
ISBN: 0-7923-9957-9

MULTIMEDIA INFORMATION SYSTEMS, by Marios C. Angelides and Schahram Dustdar
ISBN: 0-7923-9915-3

MOTION ESTIMATION ALGORITHMS FOR VIDEO COMPRESSION, by Borko Furht, Joshua Greenberg and Raymond Westwater
ISBN: 0-7923-9793-2

VIDEO DATA COMPRESSION FOR MULTIMEDIA COMPUTING, edited by Hua Harry Li, Shan Sun, Haluk Derin
ISBN: 0-7923-9790-8

REAL-TIME VIDEO COMPRESSION: *Techniques and Algorithms*, by Raymond Westwater and Borko Furht
ISBN: 0-7923-9787-8

MULTIMEDIA DATABASE MANAGEMENT SYSTEMS, by B. Prabhakaran
ISBN: 0-7923-9784-3

MULTIMEDIA TOOLS AND APPLICATIONS, edited by Borko Furht
ISBN: 0-7923-9721-5

MULTIMEDIA SYSTEMS AND TECHNIQUES, edited by Borko Furht
ISBN: 0-7923-9683-9

Cellular Automata Transforms

Theory and Applications in Multimedia Compression, Encryption, and Modeling

by

Olu Lafe

KLUWER ACADEMIC PUBLISHERS
Boston / Dordrecht / London

Distributors for North, Central and South America:
Kluwer Academic Publishers
101 Philip Drive
Assinippi Park
Norwell, Massachusetts 02061 USA
Telephone (781) 871-6600
Fax (781) 681-9045
E-Mail <kluwer@wkap.com>

Distributors for all other countries:
Kluwer Academic Publishers Group
Distribution Centre
Post Office Box 322
3300 AH Dordrecht, THE NETHERLANDS
Telephone 31 78 6392 392
Fax 31 78 6546 474
E-Mail <services@wkap.nl>

Electronic Services <http://www.wkap.nl>

Library of Congress Cataloging-in-Publication Data

Lafe, Olu, 1951-
Cellular automata transforms : theory and applications in multimedia compression, encryption and modeling / Olu Lafe.
p. cm. – (Multimedia systems and applications series ; mmsa16)
Includes bibliographical references and index.
ISBN 0-7923-7857-1 (alk.paper)
1. Cellular automata. I. Title. II. Multimedia systems and applications ; mmsa16.
QA267.5.C45 L34 2000
511.3—dc21 00-038635

Printed on acid-free paper.

Printed in the United States of America

To my wife Idowu

Contents

Preface

This book is a product of a personal research odyssey that started in the late 1980s. From the word go, my interest in *Cellular Automata* exceeded the mere academic. As an engineer my thoughts have centered on how to put these fascinating dynamical systems into serious use. Since the pioneering thoughts of *von Neumann* in the 1940s and those of *Ulam* and *von Neumann* in the 1950s, work on *Cellular Automata* has ranged from Conway's *Life* (familiar with many cellular automata enthusiasts) to the *Lattice Gas Models* (popular with practitioners of digital physics).

In *Cellular Automata Transforms (CAT)* we have found a solid approach to use *Cellular Automata* for a variety of mathematical, physical, engineering, and general modeling applications. The characteristics of these transforms are truly amazing. Some of the building blocks associated with *Cellular Automata Transforms* exhibit characteristics not uncommon with such transform techniques like wavelets. Others have features that allow for the self-generation of functions. Another class is similar to such unitary transforms like Haar, Walsh and Hadamard. Above all, *Cellular Automata Transforms* possess an efficient data encoding capability that rivals that of the *Karhunen-Loeve Transform*, believed by many to be the optimal transform in an information packing sense. This explains why massive compression of data can be achieved using a certain family of *Cellular Automata*.

In *Cellular Automata Transforms* we have a robust way of generating billions of fascinating mathematical transform bases. These information building blocks can be adapted to the peculiarities of a given problem. For example, in digitized image compression, we choose those bases that maximize the number of zeroes in the cellular automata transform coefficients. The transforms also provide a more direct way of using *Cellular Automata* for modeling (*e.g.*, physics, chemistry, biology, ecology, economics, etc.).

One fascinating nature of *Cellular Automata* is the *Boolean* or integer-based character of the underlying computational process. The immediate consequence is a tremendous encoding and decoding speed in CAT applications, especially when these are implemented in hardware. This book presents the foundational concepts on *Cellular Automata Transforms*. Application areas in *Multimedia Compression, Data Encryption*, and *Solution of Differential and Integral Equations* are showcased. *Cellular Automata Transforms* can be utilized the same way other traditional methods, *e.g.*, Fourier and Laplace transforms, are used for process modeling and analysis. The huge number of transform bases available is a major strength of these CA transforms. The basis functions are easy to generate from the evolving states of the cellular automata. Furthermore, the computing does not have to involve the huge array of cells such as the millions commonly employed in conventional *Lattice Gas Models*.

I believe this work is the proverbial "tip of the iceberg" as we consider the different applications of this class of dynamical systems. Cellular Automata provide us with a fantastic tool for analyzing many processes. It is truly fascinating to watch how simple neighborhood actions lead to complex emergent behavior. The greatest benefit is the discovery that those elementary rules of association can help us solve practical problems such as the need for secure and efficient data transport over a communications network. I hope you will share the same excitement I have about Cellular Automata Transforms as you read through the following pages.

Olu Lafe
Chesterland, Ohio

Acknowledgements

Some part of the work reported in this book was supported by grants from the National Science Foundation (NSF); Glenn Research Center of the National Aeronautics and Space Agency (NASA) at Lewis Field; and the United States Department of Agriculture.

I am indebted to former academic and research contemporaries (including Professor Deji Demuren of Old Dominion University, Dr. Tunde Ogunnaike of Du Pont Chemicals, and Professor Alex Cheng of University of Delaware) for lively discussions on evolutionary computing. I thank my friends Dr. Charles Mbanefo, Tom Norton, Chuck Hall, Bob and Suzanne Dodd, and Don G. Riling for their support and encouragement. Jim Sacher has been a constant cheerleader and an excellent adviser. I received intellectual contributions from computer scientists, engineers, physicists and mathematicians who were interns or continue to work at Lafe Technologies. These include Graeme Lufkin (encryption and attack methods, Java-based multifunction CAT visualization), Matt Schemmel (image compression), Bryon Jacob (user interface for CATlock encryption code, audio/video compression), Mike Gustafson (video compression), Andy Slocum (fast encryption key generation), Atila Boros (audio/video compression, network transport solutions), Alexandra Boros (rescaling of CAT bases to satisfy orthogonality and smoothness conditions), Brent Zboyoski (biometrics), Heesook Yoon (visualization of CAT bases), Dmitry Zhitnisky (CAT image compression libraries), Serhiy Golodnyak (compression speed optimization), Yevgen Vengrenyuk (CAT audio interface), Tunde Adegbola (audio compression), and Tosin Oni (synthetic audio generation). I thank Shelli Wells and Christine Kolb of Lafe Technologies who carried out a painstaking formatting and preparation of the final manuscript. I am grateful to my wife Idowu, and my loving children Tolu, Femi, Tola and Funso, who joined me in observing evolving cellular automata patterns at the early stages of the investigations.

Part I

Theory of Cellular Automata Transforms

The following chapters provide in-depth fundamentals of Cellular Automata Transforms.

Chapter 1

Introduction

1.1 What are Cellular Automata?

Cellular Automata (CA) are dynamical systems in which space and time are discrete. The cells are arranged in the form of a regular lattice structure and each must have a finite number of states. These states are updated synchronously according to a specified local rule of interaction. For example, a simple two-state, one-dimensional cellular automaton will consist of a line of cells/sites, each of which can take value *0* or *1*. Using a specified rule (usually deterministic), the values are updated synchronously in discrete time steps for all cells. With a K-state automaton, each cell can take any of the integer values between *0* and K - *1*. In general, the rule governing the evolution of the cellular automaton will encompass m sites up to a finite distance r away. We say the cellular automaton is a K-state, m-site neighborhood CA.

1.2 History of Cellular Automata

von Neumann and Ulam

The modern day understanding of cellular automata is rooted in the pioneering work of *von Neumann*. In 1948 he set out to simulate complex biological systems using the dynamics of simple interacting elements. *von Neumann* had designed the basic architecture of the first sequential computer using electronic logic devices. *Stan Ulam* joined *von Neumann* in the 1950s to develop the concept of a discrete model for natural events. Both realized the limitation of the serial computer in solving a large class of problems. The serial computer showed great potential in the solution of discretized partial differential equations governing problems in continua. However, as a general computing platform, the sequential computer is too complicated and demands an effort that is not necessarily commensurate with the complexity of a given problem. There must be another approach to computing.

von Neumann and Ulam were fascinated by the efficiency, robustness, and the prolonged survival of biological systems in often severe environments. They arrived at the concept of cellular spaces as a vehicle for carrying out discrete models of complex systems. Their basic reasoning went like this:

- Start with a simple system that possesses a finite state. The simplest system is a dual-state machine.

- The system will consist of a lattice structure with a network of small neighborhoods.

- There will be a rule of interaction, defined at the local (neighborhood) levels, which will be applied at the same time throughout the cellular space.

- The system will be allowed to evolve. The challenge is to see how the evolving states can be used as the main engine of a computing device.

There is a rich collection of historical notes and references on cellular automata in a number of publications by Tommaso Toffoli and Norman Margolus (See, *e.g.*, *Toffoli* [1984a&b]; *Toffoli & Margolus* [1987]; *Toffoli & Margolus* [1990]; *Toffoli & Margolus* [1991]; *Toffoli* [1994]; *Toffoli* [1995]; *Toffoli & Margolus* [1996]; and *Toffoli* [1998].)

A simple dual-state, one-dimensional cellular space is depicted in Figure 1.1. A given node can either be **on** (assigned a state value *1*) or **off** (value *0*). The closest nodes to any given node are those to its immediate left and right. In that case, we can have a local neighborhood of three cells. The state of a node at time $t + 1$ will be determined by the states of the cells within its neighborhood at time t.

Figure 1.1 *One-dimensional cellular space*

Another structure is that of a square lattice depicted in Figure 1.2. The intersection of the squares form the nodes of the automata. The closest nodes to a given node are the four to the immediate North, South, West, and East, moving along the lines connecting the nodes.

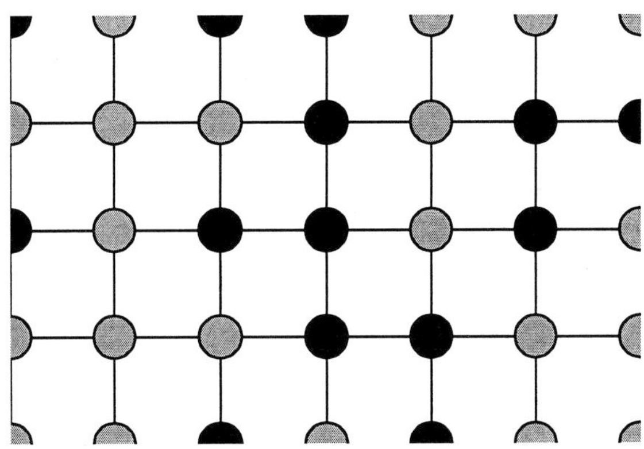

Figure 1.2 *Two-dimensional square lattice cellular space*

The specified node, with its four nearest neighbors, form the *von Neumann* neighborhood. Again, the state of the given node at time $t + 1$ will be determined from the states of the nodes within its neighborhood at time t. The rule of evolution is applied to all nodes (and their associated neighborhood) at the same time. A cellular space drawn from a hexagonal lattice is shown in Figure 1.3. In this case, a possible neighborhood is that of a node and its six closest neighbors.

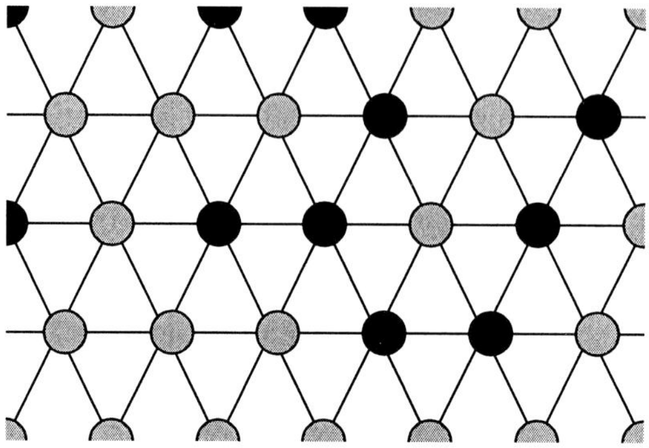

Figure 1.3 *Two-dimensional hexagonal cellular space*

Stan Ulam continued, into the 1960s, to study pattern development in cellular automata (*Ulam* [1962]). However, major progress in the understanding of CA was not realized until almost a decade later.

The Game of Life

Popular interest in cellular automata was heavily generated as a result of the work of *John Conway* who invented the *Game of Life*. Conway's *Life* presents an excellent tool for simulating biological systems. Several investigators have also looked into the universal computing properties of the *Game of Life*. *Berlekamp, Conway,* and *Guy* [1982] presented the proof that the *Game of Life* can perform universal computation. The logical gates **AND, OR, NOT** are sufficient for all logical functions. In the *Game of Life,* universal computation can be achieved by arranging interaction laws that form the basic logical gates (*Poundstone* [1985]; and *Langton* [1986]).

1.3 Multi-State CA Example

An interesting, but imperfect, example of a multi-state cellular automata is the traffic light system of a well laid-out city. The road system consists of a network of intersecting two-lane carriage ways

1.3 Multi-State CA Example

An interesting, but imperfect, example of a multi-state cellular automata is the traffic light system of a well laid-out city. The road system consists of a network of intersecting two-lane carriage ways (Figure 1.4). At each junction there are lights to direct the traffic along all four cardinal directions: North, South, West, and East. We may also have special left-turn lights. Each light will either be RED, AMBER, or GREEN. A fourth situation can be included if we consider the possibility of blinking lights. There are many levels of complexity for this real-life cellular automata field:

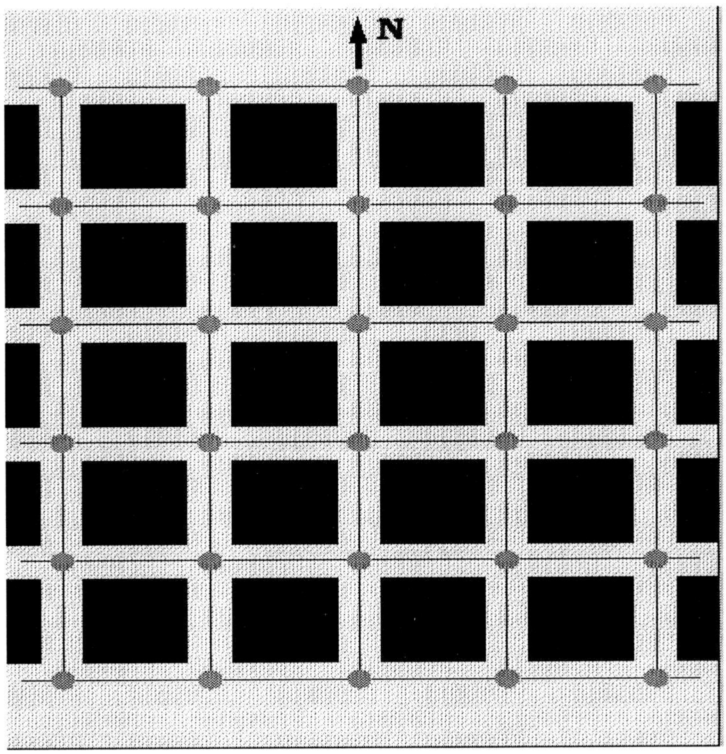

Figure 1.4 *Traffic light system as a model for multi-state cellular automata*

Table 1.1 *State permutations in a two-light junction*

State	Light 1	Light 2
0	RED	RED
1	AMBER	RED
2	GREEN	RED
3	RED	AMBER
4	RED	GREEN

As far as each junction is concerned, we have five discrete states.

2. A more complex arrangement will have four lights per junction by including the possibility of left-turn signals (two lights) to the previous arrangement. The number of states in the arrangement is summarized in Table 1.2.

Table 1.2 *State permutations in a four-light junction with two left-turn (LT) signals*

State	Light 1	Light 2	Light 3 (LT)	Light 4 (LT)
0	RED	RED	RED	RED
1	AMBER	RED	RED	RED
2	GREEN	RED	RED	RED
3	RED	AMBER	RED	RED
4	RED	GREEN	RED	RED
5	RED	RED	AMBER	RED
6	RED	RED	GREEN	RED
7	RED	RED	RED	AMBER
8	RED	RED	RED	GREEN

More complicated patterns will include independent operations of four lights for the four principal directions, left-turn signals, pedestrian crossing lights, flashing lights, etc. To obtain a representative CA field from this traffic light system, the rules governing the changing of the lights at a given junction will depend on the current state at the junction and the states of the lights in the four adjoining intersections. The lights at all intersections must be changed at the same time. An unconventional operation of these traffic-light-

More complicated patterns will include independent operations of four lights for the four principal directions, left-turn signals, pedestrian crossing lights, flashing lights, etc. To obtain a representative CA field from this traffic light system, the rules governing the changing of the lights at a given junction will depend on the current state at the junction and the states of the lights in the four adjoining intersections. The lights at all intersections must be changed at the same time. An unconventional operation of these traffic-light-based automata will include the local traffic pattern in the rule of evolution.

Having established the rule of interaction of the lights, the city traffic light system can be evolved starting from any initial configuration consistent with the operation of a normal system. Boundary conditions must be imposed at the junctions in the extremities of the city walls. The magic of cellular automata is that we can use the evolving field of these traffic lights as the basis of computation. Even a simple three-state, one light per junction is sufficient to help us model complex processes.

1.4 Cellular Automata Models

A great effort has been expended in associating cellular automata with a wide variety of phenomena, including those originating from physics, chemistry, biology, economics, and information systems. Lattice gas models make use of ideas gained from the kinetics of gases in interpreting the features of the evolving field of cellular automata. The CA rules are similar to the laws governing the collision of gas particles.

In applying CA to a physical problem, an association must be established between known physical parameters of the problem and those calculated as a result of repeated iteration (using the CA rule) from a set of initial conditions (*Gutowitz* [1990]). The process of relating particular CA rules to specific problems is not trivial. There has been great success in applying CA in the solution of several problems. Applications range from modeling physics to data encryption.

CA-based modeling offers advantages both in the representation of the underlying process (physical, chemical, biological, etc.) and in the numerical solution of the problem. We can enumerate several immediate advantages (*Cliffe et al.*, [1991]):

1. CA rules are expressible in Boolean algebra. Since no floating-point calculations are required, all bits have equal weight in a given computation. The problem of round-off errors is completely eliminated and the accuracy is limited only by the grid resolution.

2. The rules of interaction are local and simple. This makes CA models excellent candidates for massively parallel architectures and algorithms.

3. Fluid mixtures and reactions can be modeled directly.

4. The incorporation of the conditions at complicated boundaries is easily achieved.

5. Nonlinearities are a natural component of the CA model. No special treatment is required.

6. Rapid changes, such as large concentration/pressure gradients, are handled easily.

7. A fast evaluation of the parameter space and structure can be conducted without the need for extreme accuracy.

8. The underlying process (physical, chemical, biological, etc.) is easy to visualize.

9. The coding of the algorithm is relatively easy. Most CA codes consist of very few lines compared to the size involved in conventional methods.

Fluid Dynamics

Fluid dynamics is a field that has enjoyed, perhaps, the greatest attention by CA researchers. Much progress has been made. The use of CA is particularly appropriate because it is easy to visualize fluid dynamics as the interaction of many particles. These particles are engaged in simple collision patterns. The conventional method in fluid dynamics is to represent the interaction using a *continuum* model. This invariably results in partial differential equations, such as the Navier-Stokes equations. Efforts at solving these partial differential equations continue to be made by practitioners in the Computational Fluid Dynamics (CFD) field.

Cellular Automata provide a bridge between the kinetic view of fluids and the continuum model. Cellular Automata represent a serious and highly effective tool for studying the *microscopic* character of transport processes and for solving the associated *macroscopic* (continuum) equations. One of the first significant efforts at using CA for solving the Navier-Stokes equations is the work of *Frisch et al.,* [1986]. They were able to show that a class of deterministic cellular automata (or lattice-gas automata, as it is referred to in their paper) can simulate the Navier-Stokes equation. They made use of discrete Boolean elements in a hexagonal lattice CA structure.

There has been a flurry of articles on CA as models of fluid dynamics. *Wolfram* [1986] derived two-dimensional (2D) and three-dimensional (3D) continuum equations for fluids from the large-scale behavior of cellular automata. *Frisch et al.,* [1987] developed 2D (square and triangular lattices) and 3D (face-centered-hypercubic lattices) for solving Navier-Stokes equations. They utilized deterministic and non-deterministic rules. *Orszag & Yakhot* [1990] show the problems faced in using CA to model fluid dynamics when the Reynolds number is high. Other CA fluid dynamics applications include those of *Rothman & Keller* [1990], *Yakhot et al.,* [1990], *d'Humieres & Lallemand* [1987], *Boghosian* [1988], *Ernst & Shankar* [1992], *Kohring* [1992a,c], *Perera et al.,* [1992], and *Garcia-Ybarra et al.,* [1994].

A common feature of these models is the huge number of cells required for a typical simulation. For example, the work reported by *Shimomura et al.*, [1985] required 14 million cells to investigate the Kelvin-Helmholtz instability and 20 million cells to simulate flow past a cylinder. It must be acknowledged that the calculations performed at each node are relatively simple, since most required manipulation of bits. The large number of simple calculations, performed synchronously at the nodes, lends the opportunity for parallel CA computational schemes.

Transport Processes in Porous Media

The work by *Frisch et al.*, [1986] had a significant impact on porous media research by providing the vital linkage between cellular automata and the equations governing flows in continua. For example, *Rothman* [1988] and *Chen et al.*, [1991] examined the potential applications of CA methods to the fluid flows in complex porous media. *Wells et al.*, [1991] investigated CA models for simulating coupled solute transport and chemical reactions at mineral surfaces and in pore networks. Other CA applications to flow in porous media include those of *Rothman* [1990]; *Gunstensen & Rothman* [1991a]; *Kohring* [1992b]; and *Di Pietro et al.*, [1994].

Chemistry and Diffusion-Controlled Reactions

Some of the most challenging areas in modeling complex systems include chemistry and diffusion-controlled reactions. Typical chemical processes exhibit behaviors that involve the interplay of several species and multiple scale levels. Fortunately, cellular automata models are making major strides in these areas. Examples of published investigations include *Hartman & Tamayo* [1990] (chemical turbulence); *Vichniac* [1984] and *Creutz* [1986] (percolation and nucleation); *Greenberg et al.*, [1978a, b, 1980], *Canning & Droz* [1990], *Dab et al.*, [1991]; and *Weimar & Boon* [1994] (diffusion-controlled reactions).

Ecology

Ecological modeling is one of the areas in which CA application is showing great promise. There is a debate as to whether the synchronous updating of the cells may not be too limiting in developing CA ecological models (*Burke* [1994]). Published studies on the use of CA in ecological modeling include those of *Hogeweg* [1988] and *Huberman & Glance* [1993] (effects of asynchronous updating of CA cells on ecological models); *Silvertown et al.*, [1992] (grass species competition); *Nowak & May* [1992] (evolutionary games and spatial chaos); and *Green et al.*, [1982, 1985] (fire and dispersal effects on spatial patterns in forests).

Data Encryption

With the increasing digital traffic load on the information highway and the need to secure and protect the integrity of data, cryptography has emerged as a vital technology. Cellular automata provide a robust environment for developing a data encryption standard. Work in the area of CA data encryption has been intense. A number of patents have already been granted (see *e.g., Gutowitz* [1994]) and many more are in the pipeline. Available literature includes work by *Wolfram* [1985]; *Delahaye* [1991]; *Guan* [1987]; and *Gutowitz* [1993a, b, 1994]. In the patent granted to *M. Bianco and D. Reed* (U.S. Patent 5,048,086), use is made of dynamical systems to generate pseudo-random numbers that are combined in an XOR operation with the plaintext to form the encrypted message. The seed of the pseudo-random number is the encryption key. The *Wolfram* [1985] paper makes use of the cellular automata Rule 30 to generate the pseudo-random numbers. The encryption key is the initial state of the cellular automaton. In the *Guan* paper, an invertible dynamical system is used. During the encryption phase the dynamical system is run in the forward direction. Decryption involves running the inverse of the dynamical system on the encrypted message. The Gutowitz patent (U.S. Patent 5,365,589) uses irreversible dynamical systems, involving either forward and backward iteration or both, in some aspects of the encryption and decryption processes.

One primary limitation of some of the above cryptographic techniques is the complexity of the encryption and decryption processes. In the implementations that use pseudo-random numbers, the quality of the generated numbers, as pertaining to their true randomness, cannot be fully guaranteed. The ones that use forward and backward iteration of reversible and irreversible dynamical systems involve complicated mathematical operations. As will be explained below, the CAT-based cryptographic method (for which this author was given the U.S. Patent 5,677,956 on October 14, 1997) makes use of simple transform operations, which involve a huge library of cryptographic keys derived from a family of cellular automata.

The desirable properties of a good cryptographic system are:

1. Error-free encrypting/decrypting
2. Secure; tamper-proof; an ability to frustrate attempts at code-breaking
3. Error-correction capability
4. Fast operation
5. Absence of floating point computations
6. A one-to-many plaintext to ciphertext mapping capability, even for a given encryption key
7. Flexibility in accepting data of any arbitrary size

A CA-based cryptographic system possesses these qualities. Computational speed is assured by the inherent parallelism of CA computing. A floating point computation can be avoided because of the Boolean integer character of the computation and the discrete nature of the variables (*i.e.*, the states of the cells). Protection against code-breakage is offered through the plethora of reversible and non-reversible evolutionary fields possible with a given CA rule, the defined neighborhood, and assumed patterns of initial and boundary configurations.

Computing

The CA founding fathers, *von Neumann* and *Ulam*, recognized the potential of CA for universal computation. A CA-based computer is naturally a parallel processor. Discussions and ideas on the development of the CA computer have been going on for years. A 1984 description of a computer architecture based on cellular automata was presented by *Hillis* [1984] in an article that provides details on the motivations for massively parallel processors. *Despain et al.*, [1990] discussed prospects for a dedicated lattice-gas computer whose performance may exceed those of existing supercomputers 100 million times. *Margolus & Toffoli* [1990], who developed the CAM-6 simulator, discussed cellular automata machines in general. *Howard et al.*, [1992] described a three-dimensional CA processor that utilizes a massively parallel architecture.

All indications are that cellular automata computing will be key to efforts at developing new and future generations of computing devices.

1.5 Challenges in Conventional CA Modeling

To date, the route toward using cellular automata for computing or as models of physical, chemical, biological, etc., phenomena is somewhat convoluted. The main challenge is the ability to associate the given phenomenon with the evolving field of the automata.

For example, the popular way researchers in lattice-gas techniques link a particular CA rule with a continuum equation is the following, *Chen et al.*, [1991] (See Figure 1.5):

1. Create the collision rule for the lattice gas.
2. Write the microscopic equation describing the evolution of the automaton.
3. Define macroscopic parameters such as local mean density and mean momentum.
4. Introduce a statistical distribution law, *e.g.*, Fermi-Dirac distribution (*Gunstensen & Rothman* [1991b]), to expand the salient macroscopic parameters about an equilibrium.
5. Apply asymptotic analysis to find the limiting behavior of the expansions. The result is a continuum equation that is associated with the original CA rule.

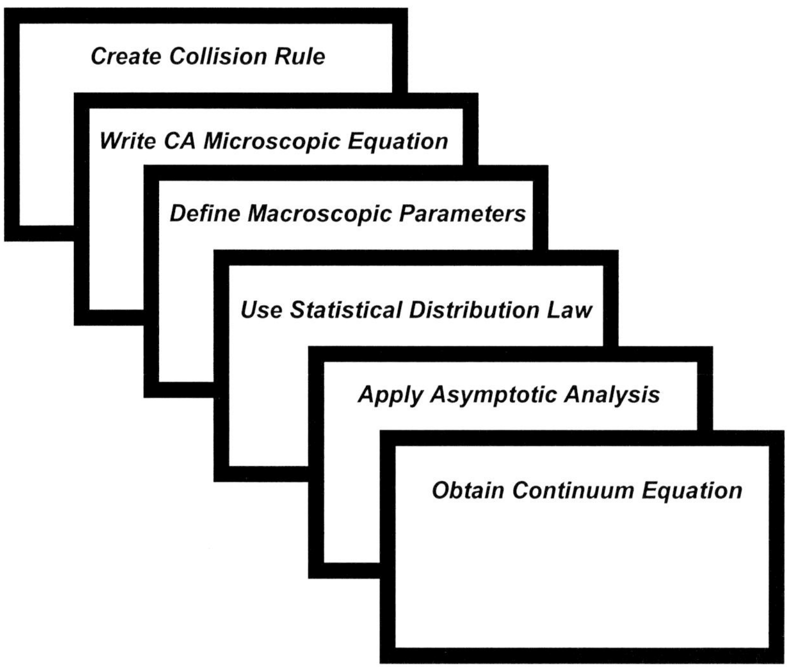

Figure 1.5 *Traditional steps in lattice-gas techniques of associating CA rules with a continuum equation*

The problem, in the above approach, is that you start with a discrete model and try to find the differential equations that govern the discrete model (*Kohring* [1992]). It is difficult to find a discrete model that is governed by the desired continuum equations. The association has been made for a few continuum equations such as the Navier-Stokes equations (*Frisch et al.*, [1986]).

1.6 Cellular Automata Transforms

Cellular Automata Transforms present a more direct way of achieving the linkage between a given phenomenon and the evolving CA field. CA transforms can be utilized in the way other transforms (*e.g.*, Fourier, Laplace, wavelets, etc.) are utilized. Cellular Automata are capable of generating billions of orthogonal, semi-orthogonal, bi-orthogonal and non-orthogonal bases. These can be adapted to the peculiarities of a given problem. Some CA bases exhibit features that look like those of established transform methods (*e.g.*, Walsh, Hadamard, Haar and wavelets). Another class can reveal the self-generating property of a data set or a function. This class can be used for:

- Compressive encoding of images
- Iterative generation of complex mathematical functions
- Multi-resolution analysis and interpolation of data

Transform Equation

Given a process described by a function f, defined in a physical space of lattice grid i, we seek basis functions (or filters) A and their associated transform coefficients c, defined in cellular automata space of lattice grid k, which allow us to write:

$$f_i = \sum_k c_k A_{ik} \tag{1.1}$$

The basis functions are related to the evolving field (or the states) of the cellular automata. Note that each point on the physical grid i has an associated basis function A (spanning the entire *CA* space). Equation (1.1) represents a mapping of the process f (in the physical domain) into c (in the cellular automata domain) using the building blocks A as transfer functions. In many applications, we seek to obtain transform coefficients c with properties not necessarily possessed by the original function f. Or the transformation process should reveal things about f not readily observed in the physical domain. For example, in data compression applications, we want the transformation to reveal the redundancies in the original data. The elements of c with insignificant or zero magnitudes reveal the degree of redundancy detected by the CA transform. In solving partial differential equations, we want the representation in equation (1.1) to automatically satisfy the governing equations and the imposed boundary/initial conditions.

The essence of *Cellular Automata Transforms* is that we can always find CA rules, with the associated gateway values, which will result in basis functions and transform coefficients with properties we desire for a given problem. The chief strength is the huge number and varied nature of the basis functions.

For example, in data compression applications, we desire, among other things:

- Small alphabet base for generating the basis functions, A
- Considerable ease (*i.e.*, computational speed) in calculating the transform coefficients, c
- Basis functions that will maximize the number of negligibly small c coefficients, while minimizing the encoding error.

On the other hand, for data encryption applications we actually seek to maximize the alphabet base used in generating the basis functions.

This will help thwart the efforts of code breakers. Furthermore, the CA transform of the data must be error-free.

In solving partial differential equations, we seek basis functions with "nice" differentiation properties. The calculation of the transform coefficients should also not require the explicit inversion of matrix equations. The last feature is provided for by using orthogonal or semi-orthogonal CA basis functions. In dealing with integral equations, we want a transformation that will result in sparse and diagonally strong coefficient matrices.

Transform Equation Related Issues

The transformation depicted in equation (1.1) raises certain critical questions:

1. How are the coefficients c to be calculated?

2. How does the physical lattice space i relate to the cellular automata lattice space k?

3. How are the basis functions A obtained from the evolving states of the cellular automata?

4. How adequate is a CA transform as a model of a particular process?

These and more fundamental questions will be answered in the ensuing chapters. Below we discuss a select group of key potential applications of the *Cellular Automata Transforms*. These include digital image coding and compression, data encryption, digital signal processing, and the solution of partial/integral equations.

1.7 Potential Applications of CAT

Digital Image Coding and Compression

Digital image coding is a broad term that encompasses a variety of image processing chores. These include image restoration, enhancement, segmentation, compression, feature extraction and pattern recognition. Cellular Automata Transforms can be used to achieve all the enumerated data processing tasks. However, the ability to use CA Transforms in the compression of digital images (including audio and video data) is particularly fascinating.

Effective data compression results when a given data is transformed by choosing CA bases that:

1. Maintain a high degree of encoding fidelity

2. Maximize the number of insignificant coefficients

3. Minimize the number of bits required to encode the significant coefficients.

The possibility of adaptive encoding makes CA Transforms excellent vehicles for extremely high compression ratios given a smart search for optimal transform bases. CA Transforms can lead to symmetric and asymmetric compression. In symmetric compression, it takes approximately the same amount of time to encode a given data as to decode it. For example, the compression of a live television broadcast will demand a symmetric compressor. An asymmetric compressor may take a shorter decoding time (Type A), or a longer decoding time (Type B) than the time taken to encode the data. In data compression for distribution purposes (*e.g.*, CD-ROM applications) or archiving purposes, the Type A asymmetric compression scheme is desirable. On the other hand, in real-time data gathering situations where the cost of data transmission may be high, and there is ample time for data post-processing, Type B should be preferred. Symmetric CA encoding makes use of relatively few transform bases. The compression achievable will obviously not be as large as that available with the adaptive CA encoder. The use of CA transform for lossy multimedia data compression is presented in detail in Chapter 4.

Data Encryption

The huge number of *Cellular Automata Transform* bases that can encode data with zero error provides an excellent means by which data can be encrypted. The encryption key consists of integral *gateway* values used in generating the CA bases. The CA transform coefficients are transmitted (or stored) in the place of the original data. Access to the data can only be achieved via the integral gateway values used in the encoding. These keys are discussed in a greater detail in later chapters. The description of the application of CAT for data encryption is presented in Chapter 5.

Digital Signal Processing

Digital signal processing is a subject that is relevant to several fields in engineering, physics and mathematics. Often an analyst needs to make inferences from data collected from a given process. These may include:

- The rate of change (derivatives) of the data at specific points
- Estimates of the data values at points not included in the original (*e.g.*, data interpolation/extrapolation, zooming on images)
- Determination of points of rapid changes (*e.g.*, edge location in images)
- Determination of key parameters unique to data (*e.g.*, pattern recognition).

Solution of Partial Differential Equations

In one approach, the solution to a given PDE is written in the form of a finite series consisting of the CA bases. The series contains coefficients whose values must be determined. The most suitable group of CA bases will result in an automatic satisfaction of the governing equation at all computational nodes, regardless of the nature of the imposed boundary and initial conditions. In reality, we can only hope to minimize the error at these nodes.

The most challenging aspects in this CAT-based approach to solving PDEs are:

1. Accurate differentiation of the CA bases A.

2. Determination of the coefficients c. When orthogonal CA bases are used, the calculation of the coefficients is quite straightforward. In other cases, especially when the governing equations are nonlinear, an elaborate scheme may be required to determine the coefficients.

Solution of Integral Equations

The orthogonal property of a large number of CA transform bases and their capability to transform data with relatively few insignificant coefficients, provide an excellent platform for solving integral equations. The kernels of integral equations are transformed into a CA space in which the ensuing matrices are sparse, banded, and possess robust inversion properties. The huge number of CA bases permits ample choice of transforms to suit the characteristics of a given integral equation.

The application of CAT in solving differential and integral equations is explored in detail in Chapter 6.

Chapter 2

Cellular Automata Transforms

2.1 Nomenclature

Consider a three-site neighborhood, dual-state, one-dimensional CA. The state of each cell is given by the Boolean variable a. When the state is **on**, $a=1$. Otherwise it is **off** and $a=0$. The quantity a_{it} represents the state (Boolean) of the i-th cell, at discrete time t, whose two neighbors are in the following states: a_{i-1t}, a_{i+1t}. In general, we seek a rule that will be used to synchronously calculate the state a_{it+1} from the state of the cells in the neighborhood at the t-th time level (Figure 2.1). The cellular automaton evolution is expressible in the form:

$$a_{it+1} = F(a_{i-1t}, a_{it}, a_{i+1t}) \tag{2.1}$$

where F is a Boolean function defining the rule.

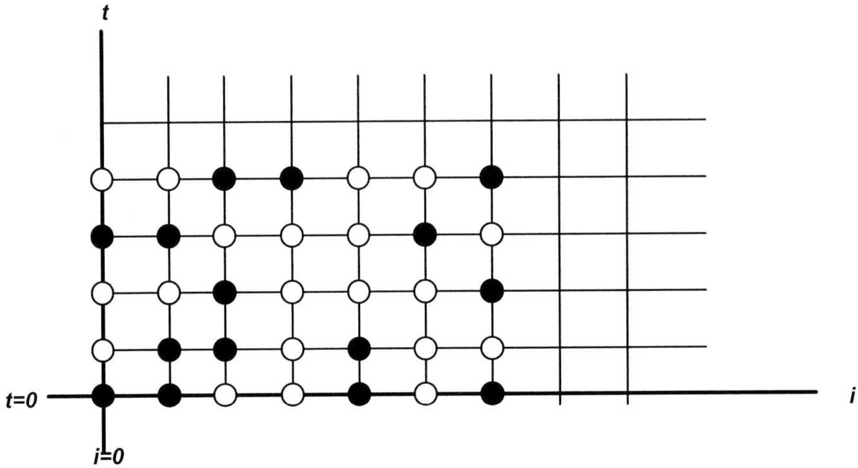

Figure 2.1 *Dual-state cellular automata lattice*

Multi-State Multi-Dimensional Rules in Cellular Automata Transforms

The number of cellular automata rules can be astronomical even for a modest lattice space, neighborhood size, and CA state. Therefore, in order to develop practical applications, a system must be developed for addressing a subset of this infinitely large universe of CA rules. Consider, for example, a *K*-state *N*-node cellular automaton with $m=2r+1$ points per neighborhood. Hence, in each neighborhood, if we choose a numbering system that is localized to each neighborhood, we have the following representing the states of the cells at time *t*: a_{it} $(i=0,1,2,3,\ldots m-1)$. We define the rule of evolution of a cellular automaton by using a vector of integers W_j $(j=0,1,2,3,\ldots 2^m)$ such that

$$a_{(r)(t+1)} = \left(\sum_{j=0}^{2^m-2} W_j \alpha_j + W_{2^m-1} \right)^{W_{2^m}} \bmod K$$

where $0 \le W_j < K$ and α_j are made up of the permutations of the states of the cells in the neighborhood. To illustrate these permutations, consider a three-site neighborhood, one-dimensional CA. Since $m=3$, there are $2^3=8$ integer W values. The states of the cells are (from left to right) a_{0k}, a_{1k}, a_{2k} at time t. The state of the middle cell at time $t+1$ is:

$$a_{1(t+1)} = \left(W_0 a_{0t} + W_1 a_{1t} + W_2 a_{2t} + W_3 a_{0t} a_{1t} + W_4 a_{1t} a_{2t} + W_5 a_{2t} a_{0t} + W_6 a_{0t} a_{1t} a_{2t} + W_7\right)^{W_8} \bmod K$$

Hence, each set of W_j results in a given rule of evolution. The chief advantage of the above rule-numbering scheme is that the number of integers is a function of the neighborhood size; it is independent of the maximum state, K, and the shape/size of the lattice. We refer to this rule system as the W-set or W-Rule throughout this book.

A sample C code is shown in Appendix A for evolving one-dimensional cellular automata using a reduced set ($W^{2m}=1$) of the W-set rule system.

Wolfram Dual-State One-Dimensional Rules

Wolfram [1983] developed a set of simple rules for describing dual-state one-dimensional cellular automata. There are $2^3=8$ possible configurations for each neighborhood in a dual-state three-site neighborhood automaton. These are:

CONFIGURATION	BOOLEAN VALUE, C
111	C_0
110	C_1
101	C_2
100	C_3
011	C_4
010	C_5
001	C_6
000	C_7

in which C_n is the Boolean value generated by the rule given the n-th configuration. There are 2^8 rules for the two-state/three-site CA. The Wolfram Rule convention assigns the integer R to the rule generating the function F such that:

$$R = \sum_{n=0}^{7} C_n 2^n$$

Hence, R takes on the value between 0 and 255 for a two-state/three-site CA.

The W-set (with $W_8=1$), which generates some of the dual-state, three-site neighborhood Wolfram Rule, is shown in the table below.

Table 2.1 *Relationship between some Wolfram Rules and W-set*

Wolfram Rule	W_0	W_1	W_2	W_3	W_4	W_5	W_6	W_7
252	1	1	0	1	0	0	0	0
195	0	1	1	0	0	0	0	1
127	0	0	0	0	0	0	1	1
16	1	0	1	1	1	0	0	0

To translate a Wolfram Rule with the binary representation $x_7x_6x_5x_4x_3x_2x_1x_0$ to the W-set, the following relationships can be used:

$$\begin{aligned}
W_7 &= x_0 \\
W_6 &= x_1 - W_7 \\
W_5 &= x_2 - W_7 \\
W_4 &= x_3 - W_5 - W_6 - W_7 \\
W_3 &= x_4 - W_7 \\
W_2 &= x_5 - W_5 - W_6 - W_7 \\
W_1 &= x_6 - W_3 - W_5 - W_7 \\
W_0 &= x_7 - W_1 - W_2 - W_3 - W_4 - W_5 - W_6 - W_7
\end{aligned}$$

We use the Wolfram Rule nomenclature when we are dealing with simple dual-state, 3-site neighborhood, one-dimensional cellular automata. In all other cases we use the W-set system. In general, for a K-state/m-site CA, there are K^{K^m} rules and the evolution is expressible in the form:

$$a_{it+1} = F(a_{i-rt}, a_{i-r+t}, a_{i+rt}) \tag{2.2}$$

If there are N cells in the entire one-dimensional space, we have a total of K^N possible initial configurations with which to start the evolution of the CA. Furthermore, if the CA is run over T discrete time steps, the number of boundary (left and right) configurations[1] possible is K^{2T}. Since there are K^{K^m} rules, the number of ways, N_T, we can evolve a k-state/m-site/N-cells CA to over T time steps is of the order:

$$K^{K^m+N+2T}$$

As summarized in Table 2.2 (K=2, m=3, T=N) the magnitude of this number is astronomical even for the most elementary cellular automata.

Table 2.2 *Number of ways of evolving one-dimensional CA*

N	8	16	32	64	128	256
N_T	2^{32}	2^{56}	2^{104}	2^{200}	2^{392}	2^{776}

For reasons that will become more evident later, the one-dimensional CA provides sufficient foundation for transforming data in any number of dimensions. Excellent multi-dimensional bases, with desirable properties, can be generated from their one-dimensional counterparts.

[1] In some implementations it is common to derive the boundary configurations from the evolving field itself by imposing a periodic (or cyclic) condition.

Two-Dimensional Cellular Space

Square Lattice

The simplest neighborhood is the *von Neumann* type consisting of the node and its four closest neighbors as depicted in Figure 1.2. In that case, a K-state automaton can be evolved with K^{K^5} rules. For a dual-state CA there are 4,294,967,296 (*i.e.*, 2^{32}) rules. If all eight closest nodes are included in the neighborhood, there are K^{K^9} rules. That is 2^{512} in the dual-state mode. For an *NM* total cell space, the square lattice has K^{N+M} different patterns of initial configuration. If the CA is evolved over T discrete time steps, the number of boundary configuration patterns is $K^{2(N+M+T)}$. Table 2.3 shows the number of ways $N_T = K^{K^m+3(N+M)+2T}$ of evolving the dual-state, two-dimensional CA with the *von Neumann* neighborhood of five cells when M=T=N.

Table 2.3 *Number of ways of evolving two-dimensional CA*

N	8	16	32	64	128	256
N_T	2^{96}	2^{160}	2^{288}	2^{544}	2^{1056}	2^{2080}

Hexagonal Lattice

The seven-cell neighborhood, see Figure 1.3, consisting of a node and its six nearest neighbors is a typical configuration for the hexagonal lattice. This cellular space has been used in the lattice-gas hydrodynamics models (*e.g.*, *Frisch* [1987]). The number of rules in the dual-state mode is 2^{128}. The number of ways to evolve the CA can be calculated as done above for one- and two-dimensional lattice spaces.

2.2 Cellular Automata Transform Bases

Given a data f in a D dimensional space measured by the independent discrete variable i, we seek a transformation in the form:

$$f_i = \sum_k c_k A_{ik} \tag{2.3}$$

where A are cellular automata transform bases, k is a vector (defined in D) of non-negative integers, while c are transform coefficients whose values are obtained from the inverse transform:

$$c_k = \sum_i f_i B_{ik} \tag{2.4}$$

in which the bases B are the inverse of A.

When the bases A are orthogonal, the number of transform coefficients is equal to that in the original data f. Furthermore, orthogonal transformation offers considerable simplicity in the calculation of the coefficients. From the point-of-view of solving PDEs, data encoding, and general digital signal processing applications, orthogonal transforms are preferable on account of their computational efficiency and elegance. *Non-orthogonal* CA bases, which are important for self-generating transform schemes, can easily be constructed using the same tools we will outline in this chapter.

The forward and inverse transform bases A and B are generated from the evolving states a of the cellular automata. Below is the outline of how these bases are generated.

Classification of CA Transforms

A given CA transform is characterized by one (or a combination) of the following features:

1. The method used in calculating the bases from the evolving states of cellular automata.

2. The orthogonality or non-orthogonality of the basis functions.

3. The method used in calculating the transform coefficients. Orthogonal transformation is the easiest. In self-generating transformation, we exploit any self-similarity in the data by using transform coefficients, which are approximations of the function being transformed.

The simplest bases are those with coefficients (1,-1) and are usually derived from dual-state cellular automata. Some bases are generated from the instantaneous point density of the evolving field of the cellular automata. Other basis functions are generated from a multiple-cell-averaged density of the evolving automata.

Construction of CA Bases

One-dimensional (D=1) cellular spaces offer the simplest environment for generating CA transform bases. They offer several advantages, including:

- A manageable alphabet base (see N_T in Table 2.2) for small neighborhood size m and maximum state K. This is a strong advantage in data compression applications because of the small number of coefficients required to describe the rules.
- The possibility of generating higher-dimensional bases from combinations of the one-dimensional bases.
- The excellent knowledge base of one-dimensional cellular automata.

In a 1D space our goal is to generate the transform basis function:

$$A = A_{ik} \qquad i,k = 0,1,2,\ldots N-1$$

from a field of L cells evolved for T time steps. Therefore consider the data sequence $f_i(i = 0,1,2,\ldots N-1)$. We write:

$$f_i = \sum_{k=0}^{N-1} c_k A_{ik} \qquad i,k = 0,1,2,\ldots N-1 \tag{2.5}$$

in which c_k are the transform coefficients. There are infinite ways by which A_{ik} can be expressed as a function of the evolving field of the cellular automata $a = a_{it}$, (i=0,1,2,...L - 1; t=0,1,2,...T-1). A few of these are enumerated below. The simplest way of generating the bases is to evolve N cells over N time steps (Figure 2.2). That is L=T=N. This results in N^2 coefficients from which the building blocks A_{ik} can be derived. We call this the Class I Scheme.

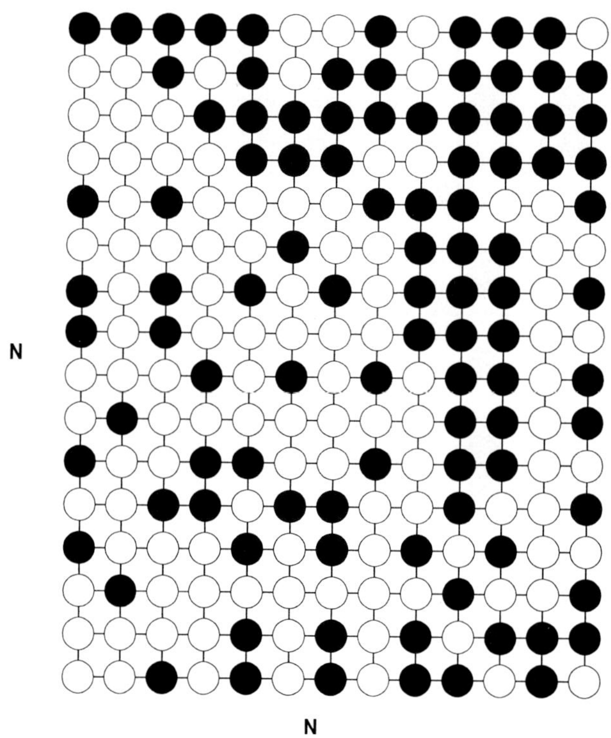

Figure 2.2 *Cell arrangements in Class I Scheme. The bottom base states form the initial configuration. The N cells are evolved over N time steps.*

A more universal approach, the Class II Scheme, selects $L=N^2$ (*i.e.*, the number of coefficients to be derived) and makes the evolution time T independent of the size of the basis function (Figure 2.3). One major advantage of the latter approach is the flexibility to tie the basis precision to the evolution time T.

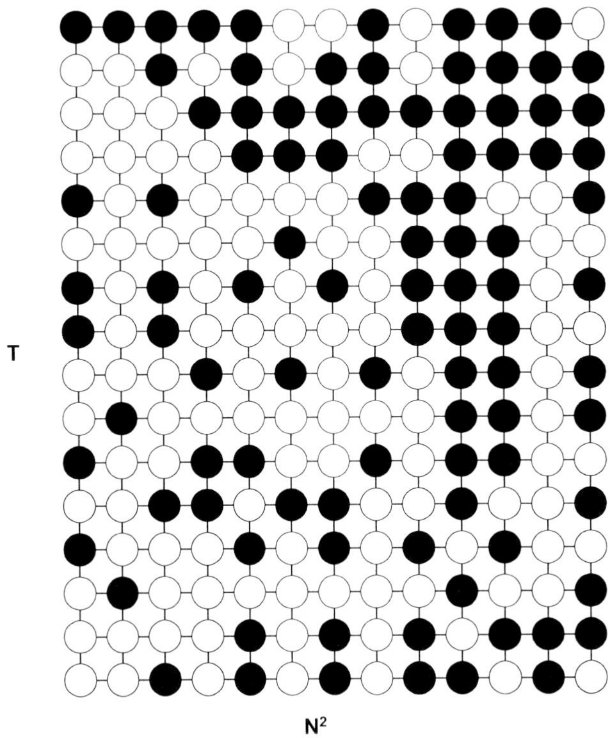

Figure 2.3 *Cell arrangements in Class II Scheme. The bottom base states (with N² cells) form the initial configuration. The N²cells are evolved over T time steps.*

Class I Scheme

When the N cells are evolved over N time steps, we obtain N^2 integer coefficients:

$$a = a_{it}, \quad i,t = 0,1,2,\ldots N-1$$

which are the states of the cellular automata including the initial configuration. A few basis types belonging to this group include:

- **Type 1:** $A_{ik} = \alpha + \beta a_{ik}$

  where a_{ik} is the state of the CA at the node i at time $t=k$ while α and β are constants.

- **Type 2:** $A_{ik} = \alpha + \beta a_{ik} a_{ki}$

- **Type 3:** $A_{ik} = \mu_{ik} \mu_{ki}$

$$\mu_{ik} = \alpha \sum_{l=i-n_w}^{i+n_w} a_{l_w k}$$

$$l_w = \begin{cases} 1 & if \;\; 0 \le l \le N-1 \\ l+N & if \;\; l < 0 \\ l-N & if \;\; l > N-1 \end{cases}$$

  where $1 \le n_W \le N-2$; implying there are $(N-2)$ different ways of generating Type 3 bases. When used as a group transform, the decomposition will be in the form:

$$f_i = \sum_{n_w=1}^{(N-2)} \sum_{k=0}^{N-1} c_{n_w k} A_{n_w ik}$$

- **Type 4:** $A_{ik} = \rho_{ik} \rho_{ki}$

$$\rho_{ik} = \alpha a_{ik} + \beta + \rho_{ik-1}$$

$$\rho_{i0} = \alpha a_{i0} + \beta$$

- **Type 5:**

$$A_{ik} = \rho_{ik}\rho_{ki}$$

$$\rho_{ik} = \alpha a_{ik} + \beta + \rho_{ik-1} \quad \text{mod Lw}$$

$$\rho_{i0} = \alpha a_{i0} + \beta \quad \text{mod } L_w$$

in which $L_w \geq 2$ is an integer. There are as many ways of generating Type 5 bases as are the selection of L_w.

- **Type 6:**

$$A_{ik} = \mu_{ik}\mu_{ki}$$

$$\mu_{ik} = \mu_{ik-1} + a_{ik} / \sigma_k$$

$$\mu_{i0} = a_{i0} / \sigma_{i0}$$

$$\sigma_k = 1 + \sum_{i=0}^{N-1} a_{ik}$$

- **Type 7:** These are constructed the same way Types 1-6 are constructed but with a decimation mask applied to some coefficients. For example, let $N_W < N$ be a "window size." Then we set:

$$A_{ik} = 0 \ \text{ when}$$

$$(N_w + i) \bmod N < k < (N + i) \bmod N$$

Class II Scheme

We showcase two types of basis functions under this scheme:

- **Type 1:** $$A_{ik} = \alpha + \beta \sum_{t=0}^{T-1} a_{(k+iN)(T-1-t)} / K^t$$

in which K is the maximum state of the automaton.

- **Type 2:** $$A_{ik} = \sum_{t=0}^{T-1} \{a_{(k+iN)(T-1-t)} - \beta\}$$

Orthogonal CA Bases

In most applications we desire to have basis functions that are orthogonal. That is, we want bases A_{ik} to satisfy:

$$\sum_{i=0}^{N-1} A_{ik} A_{il} = \begin{cases} \lambda_k & k = l \\ 0 & k \neq l \end{cases} \tag{2.6}$$

where $\lambda_k \left(k = 0,1,\cdots N-1\right)$ are coefficients. The transform coefficients are easily computed as:

$$c_k = \frac{1}{\lambda_k} \sum_{i=0}^{N-1} f_i A_{ik}$$

That is, the inverse transform bases are:

$$B_{ik} = \frac{A_{ik}}{\lambda_k}$$

The bases are orthonormal when $\lambda_k=1$ for all k. Orthonormal bases, A' and B' are easily obtained from the orthogonal functions A and B via the following rescaling:

$$A'_{ik} = A_{ik} \Big/ \sqrt{\lambda_k}$$

$$B'_{ik} = \sqrt{\lambda_k}\, B_{ik}$$

A limited set of orthogonal CA bases is symmetric: $A_{ik}=A_{ki}$. The symmetry property can be exploited in accelerating the CA transform process.

The basis functions calculated from the CA states will generally not be orthogonal. There are simple normalization and scaling schemes that can be utilized to make these orthogonal and also satisfy additional conditions (*e.g.*, smoothness of reconstructed data) that may be required for a given problem.

Two-Dimensional Bases

In a 2D square space consisting of N x N cells, the transform base $\mathrm{A} = A_{ijkl}, (i, j, k, l = 0,1,\ldots N-1)$. For the data sequence, $f_{ij}(i, j = 0,1,2,\ldots N-1)$, we can write:

$$f_{ij} = \sum_{k=0}^{N-1}\sum_{l=0}^{N-1} c_{kl} A_{ijkl} \quad i, j = 0,1,2,\ldots N-1 \tag{2.7}$$

in which c_{kl} are the transform coefficients.

There are two approaches for generating two-dimensional CA transform bases:

1. Using the evolving states derived from two-dimensional cellular spaces. Here, A_{ijkl} are calculated from

   $$a = a_{ijt}(i = 0, 1, 2, \ldots L_1\text{-}1; j = 0, 1, 2, \ldots L_2\text{-}1; t = 0, 1, 2, \ldots T\text{-}1)$$

2. Using the products of one-dimensional bases.

Bases from 2D Square Cellular Space

We can use the two schemes earlier for generating 2D basis functions. A selection of Scheme I derived bases are presented below.

- **Type 1:** $A_{ijkl} = \alpha + \beta \prod_{mnp} a_{mnp}$

  where a_{mnp} is the state of the CA at the node (m, n) at time $t=p$ and the term $\prod_{mnp}(\cdot)_{mnp}$ is a product series with the indices:

$$\begin{cases} m = i, j, k, l \neq n \neq p \\ n = i, j, k, l \neq m \neq p \\ p = i, j, k, l \neq m \neq n \end{cases}$$

- **Type 2:** $A_{ijkl} = \alpha + \beta a_{ijk} a_{ijl}$
- **Type 3:** $A_{ijkl} = \prod_{mnp} \mu_{mnp}$

$$\rho_{\mathrm{mnp}} = \beta \sum_{u=m-n_w}^{m+n_w} \sum_{v=n-n_w}^{n+n_w} a_{u_w v_w k} a_{u_w v_w l}$$

$$u_w = \begin{cases} u & \text{if } \; 0 \leq u \leq N-1 \\ u+N & \text{if } \; u < 0 \\ u-N & \text{if } \; u > N-1 \end{cases}$$

$$v_w = \begin{cases} v & \text{if } \; 0 \leq v \leq N-1 \\ v+N & \text{if } \; v < 0 \\ v-N & \text{if } \; v > N-1 \end{cases}$$

where $1 \leq n_w \leq N-2$.

- **Type 4:** $A_{ijkl} = \prod_{mnp} \rho_{mnp}$

$$\rho_{mnp} = \alpha a_{mnp} + \beta + \rho_{mnp-1}$$

$$\rho_{mn0} = \alpha a_{mn0} + \beta$$

- **Type 5:** $A_{ijkl} = \prod_{mnp} \rho_{mnp}$

$$\rho_{mnp} = \alpha a_{mnp} + \beta + \rho_{mnp-1} \mod L_w$$

$$\rho_{mn0} = \alpha a_{mn0} + \beta \mod L_w$$

in which $L_W \geq 2$ is an integer.

- **Type 6:**

$$A_{ijkl} = \prod_{mnp} \mu_{mnp}$$

$$\mu_{mnp} = \mu_{mnp-1} + a_{mnp} / \sigma_p$$

$$\mu_{mn0} = a_{i0} / \sigma_0$$

$$\sigma_p = 1 + \sum_{i=0}^{N-1} \sum_{j=0}^{N-1} a_{ijp}$$

- **Type 7:** Like its one-dimensional counterpart, these are also constructed the same way Types 1-6 are built, but with a decimation mask applied to some coefficients outside a specified window area. With the window size $N_{wi} \times N_{wj}$ such that $N_{wi} < N$ and $N_{wj} < N$ we set $A_{ijkl} = 0$ when:

$$(N_{wi} + i) \bmod N < k < (N + i) \bmod N \quad \text{and}$$

$$(N_{wj} + j) \bmod N < l < (N + j) \bmod N$$

Two-Dimensional Bases Derived from Products of One-Dimensional Functions

The bases A_{ijkl} are derived from the one-dimensional types in the form:

$$A_{ijkl} = A_{ik} A_{jl}$$

There are as many canonical 2D bases as are permutations of 1D bases.

One interesting 2D basis function, which will call **Type 8**, are derived from the evolving one-dimensional automata as:

$$A_{ijkl} = L_w \left\{ \left(a_{ik} a_{ki} + a_{jl} a_{lj} \right) \bmod L_w \right\} - \left(L_w - 1 \right) \tag{2.8}$$

where $L_w \geq 2$ is the number of states of the automaton.

Multi-Dimensional CA Bases

The generation of multi-dimensional bases follows the style enumerated above for 2D bases. Use can either be made of multi-dimensional automata or products of 1D bases. On account of the small alphabet base, those derived from products of the one-dimensional bases have particular advantages in multimedia compression.

2.3 Important Keys in CA Transforms

The following are the 10 most important keys for carrying out Cellular Automata Transforms (see Table 2.4):

1. The *rule* of interaction of the cells within a defined neighborhood.

2. The number of *states* (≥ 2) allowed for each cell.

3. The number of *cells* within each *neighborhood.*

4. The total number of *cells* in the entire *lattice.*

5. The initial configuration of the cells.

6. The boundary configuration of the cells.

7. The *form* (*e.g.*, one-dimensional, square, hexagonal, etc.) of the cellular *space/structure*.

8. The *dimensionality* of the cellular space.

9. The *type* (*e.g.*, standard orthogonal, progressive orthogonal, non-orthogonal, self-generating) of the CA *transform.*

10. The *type* (*e.g.*, Types 1 through 8 above) of the CA *bases.*

The selection of the keys will govern the properties of the ensuing transform. In most applications, some of the keys (*e.g.*, 3, 4, 7) are fixed. The remaining keys are allowed to vary.

Table 2.4 *Description of CA gateway keys*

KEY	DESCRIPTION
1	CA Rule of Interaction
2	Maximum Number of States Per Cell
3	Number of Cells Per Neighborhood
4	Number of Cells in Lattice
5	Initial Configuration
6	Boundary Configuration
7	Geometric Structure of CA Space
8	Dimensionality of CA Space
9	Type of CA Transform
10	Type of CA Transform Basis Functions

2.4 Non-Overlapping and Overlapping CAT Filters

The tacit assumption in the above derivations is that the CA filters are applied in a non-overlapping manner. Hence, given data f of length L, the filters A of size N x N are applied in the form:

$$f_i = \sum_{k=0}^{N-1} c_{kj} A_{(i \bmod N)k} \tag{2.9}$$

where $i=0,1,2,\ldots L-1$ and $j=0,1,2,\ldots(L/N)-1$ is a counter for the non-overlapping segments. The transform coefficients for points belonging to a particular segment are obtained solely from data points belonging to that segment.

However, CA basis functions can be evolved as overlapping filters. In this case, if $l=N-N_l$ is the overlap, then the transform equation will be in the form:

$$f_i = \sum_{k=0}^{N-1} c_{kj} A_{(i \bmod N_l)k} \tag{2.10}$$

where $i=0,1,2,\ldots L-1$ and $j=0,1,2,\ldots(L/N_l)-1$ is the counter for overlapping segments. The condition at the end of the segment when $i > L-N$ is handled by either zero padding or the usual assumption that the data is cyclic. The use of overlapping filters allows the natural connectivity that exists in a given data to be preserved through the transform process. Overlapping filters generally produce smooth reconstructed signals even after a heavy decimation of a large number of the transform coefficients. This property is important in the compression of digital images and video signals.

2.5 CAT Sub-Band Coding

Sub-band coding is a characteristic of a large class of cellular automata transforms. Sub-band coding, which is also a feature of many existing transform techniques (*e.g.*, wavelets), allows a signal to be decomposed into both low and high frequency components. It provides a tool for conducting the multi-resolution analysis of a data sequence. Consider, for example, a one-dimensional data sequence, f_i, of length $L=2^n$, where n is an integer. We transform this data by selecting M segments of the data at a time. The resulting coefficients are sorted into two groups (Figure 2.4); those in the even location fall into one group, and the odd points in the other. The "even" group is further transformed and the resulting 2^{n-1} coefficients are sorted into two groups of even and odd located values. The odd group is added to the odd group in the first stage; and the even group is again transformed. This process continues until the residual odd and even group is of size $N/2$. The $N/2$ coefficients belonging to the odd group are added to the set of all odd-located coefficients, while the last $N/2$ even-located group coefficients form the coefficients at the coarsest level. This last group is equivalent to the low CAT frequencies of the signal. At the end of this hierarchical process we actually end up with $L = 2^n$ coefficients. To recover the original data the process is reversed: we start from the $N/2$ low frequency coefficients and $N/2$ high frequency coefficients to form N coefficients; arrange this alternately in their even and odd locations; and the resulting N coefficients are reverse transformed. The resulting N coefficients form the even parts of the next $2N$ coefficients, while the coefficients stored in the odd group form the odd portion. This process is continued until the original L data points are recovered. The above assumes the filters are non-overlapping. If overlapping filters are used the quantity N should be replaced with $N_l = N - l$, where l is the overlap.

A large class of basis functions derived from the evolving field of cellular automata naturally possess the sub-band transform character. In many others we impose the sub-band character by re-scaling the natural basis functions.

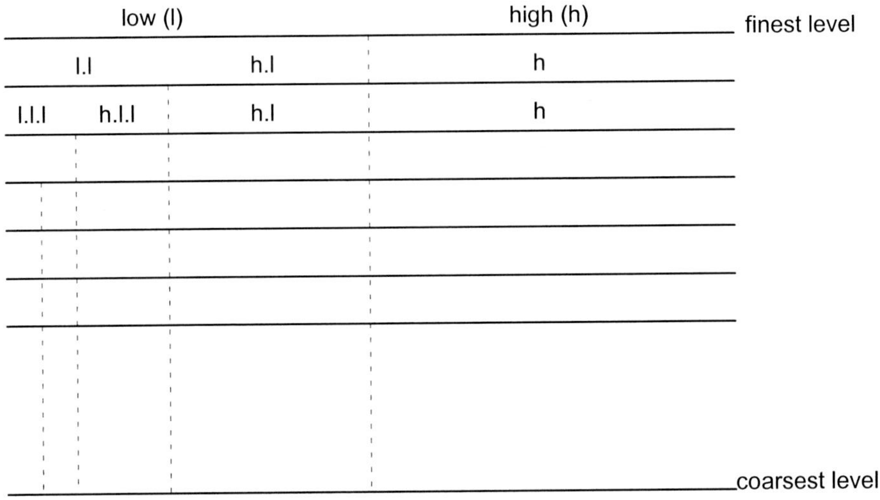

Figure 2.4 *One-dimensional, sub-band transform of a data sequence of length L. At the finest level, the transform coefficients are grouped into two equal low (l) and high (h) frequencies. The low frequencies are further transformed and regrouped into high-low and low-low frequencies each of size L/4.*

2.6 Smoothness of Sub-Band CA Basis Functions

One of the immediate consequences of sub-band coding is the possibility of imposing a degree of smoothness on the associated basis functions. We know a sub-band coder segments the data into two parts: low and high frequencies. If an infinitely smooth function is transformed using a sub-band basis function, all the high frequency coefficients should vanish. In reality we can only obtain this condition up to a specified degree. For example, a polynomial function, $f(x)=x^n$, has an n-th order smoothness because it is differentiable n times. Therefore, for the basis functions A_{ik} to be of n-order smoothness, we must demand that all the high frequency transform coefficients must vanish when the input data is up to an n-th order polynomial. That is, with $f(x)=f(i)=i^m$, we must have:

$$c_k = \sum_{i=0}^{N-1} i^m A_{ki} = 0$$
$$k = 1,3,5,\ldots; m = 0,1,2,\ldots n \tag{2.11}$$

In theory, the rules of evolution of the CA and the initial configuration can be selected such that the above conditions are satisfied. In practice, the above conditions can be obtained for a large class of CA rules by some smart re-scaling of the basis coefficients. Examples of such re-scaling will be presented in Chapter 3.

Chapter 3

Cellular Automata Bases

In this chapter, we examine the characteristics of some basis functions generated from the evolving field of cellular automata. The simplest of these bases are those derived from dual-state, three-site neighborhood CA. The basis coefficients are *(1,-1)*. More complex basis functions, satisfying additional conditions (*e.g.*, the "smoothness" and specific information-packing capability), can be evolved from multi-state, multi-neighborhood CA.

3.1 Dual-Coefficient Basis Functions

One-Dimensional Bases

An excellent example of a dual-coefficient CA basis function is the Type 2 described by:

$$A_{ik} = 2a_{ik}a_{ki} - 1 \tag{3.1}$$

where a_{ik} is the state of the CA at the node i at time $t=k$. The states are obtained from N cells evolved from a specific initial configuration for N time steps. Given:

$$f_i = \sum_{k=0}^{N-1} c_k A_{ik} \quad i = 0,1,2,\ldots N-1 \tag{3.2}$$

If A is orthogonal: $$c_k = \frac{1}{\lambda_k} \sum_{i=0}^{N-1} f_i A_{ik}$$

where $\lambda_k = \sum_{i=0}^{N-1} A_{ik}^{\ 2}$. Table 3.1 shows the coefficients for a typical orthogonal (*1,-1*) Type 2 basis function. The gateway keys used in the generation are tabulated below:

Table 3.1 *Gateway Values*

Wolfram Rule Number	11
N	8
Initial Configuration	00111111
Boundary Configuration	Cyclic
Basis Function Type	2

The cyclic boundary conditions imposed on the end sites (*i*=-1 and *i*=*N*) are of the form:

$$a_{-1k} = a_{N-1k}$$

$$a_{Nk} = a_{0k}$$

Table 3.2 *A Type 2 Basis Function*

k →	0	1	2	3	4	5	6	7
i ↓								
0	-	-	+	-	+	-	+	+
1	-	+	+	-	-	-	-	-
2	+	+	-	-	+	-	+	-
3	-	-	-	+	+	-	-	-
4	+	-	+	+	-	-	+	-
5	-	-	-	-	-	+	+	-
6	+	-	+	-	+	+	-	-
7	+	-	-	-	-	-	-	+

In spite of the similarity, these basis functions are not identical to either the Walsh (see Table 3.3) or Hadamard bases.

Table 3.3 *Walsh 8-point basis function*

k →	0	1	2	3	4	5	6	7
i ↓								
0	+	+	+	+	+	+	+	+
1	+	+	+	+	-	-	-	-
2	+	+	-	-	-	-	+	+
3	+	+	-	-	+	+	-	-
4	+	-	-	+	+	-	-	+
5	+	-	-	+	-	+	+	-
6	+	-	+	-	-	+	-	+
7	+	-	+	-	+	-	+	-

The CA gateway keys for other orthogonal dual-coefficient Type 2 basis functions using cyclic boundary conditions are shown in Table 3.4.

Table 3.4 *Gateway values for other dual-coefficient bases*

Wolfram Rule Number	*Initial Configuration*	*N*
14	11010100	8
15	10010000	8
43	01111110	8
47	11110011	8
142	01101010	8
143	01011101	8
158	11011000	8
159	01011101	8
15	0110110001011111	16
142	1101011010100101	16

Table 3.5 *Basis functions of Dual-Coefficient transform generated with a 16-point Wolfram Rule 15 CA*

k →	**0**	**1**	**2**	**3**	**4**	**5**	**6**	**7**	**8**	**9**	**10**	**11**	**12**	**13**	**14**	**15**
i ↓																
0	-	-	+	-	+	-	-	-	-	+	-	-	+	+	+	-
1	-	+	-	-	+	-	-	+	+	+	-	+	-	-	-	-
2	+	-	-	-	+	-	+	-	-	-	-	+	-	-	+	+
3	-	-	-	+	-	-	+	-	-	+	+	+	-	+	-	-
4	+	+	+	-	-	-	+	-	+	-	-	-	-	+	-	-
5	-	-	-	-	-	+	-	-	+	-	-	+	+	+	-	+
6	-	-	+	+	+	-	-	-	+	-	+	-	-	-	-	+
7	-	+	-	-	-	-	-	+	-	-	+	-	-	+	+	+
8	-	+	-	-	+	+	+	-	-	-	+	-	+	-	-	-
9	+	+	-	+	-	-	-	-	-	+	-	-	+	-	-	+
10	-	-	-	+	-	-	+	+	+	-	-	-	+	-	+	-
11	-	+	+	+	-	+	-	-	-	-	-	+	-	-	+	-
12	+	-	-	-	-	+	-	-	+	+	+	-	-	-	+	-
13	+	-	-	+	+	+	-	+	-	-	-	-	-	+	-	-
14	+	-	+	-	-	-	-	+	-	-	+	+	+	-	-	-
15	-	-	+	-	-	+	+	+	-	+	-	-	-	-	-	+

A graphical display of Table 3.5 is shown in Figure 3.1.

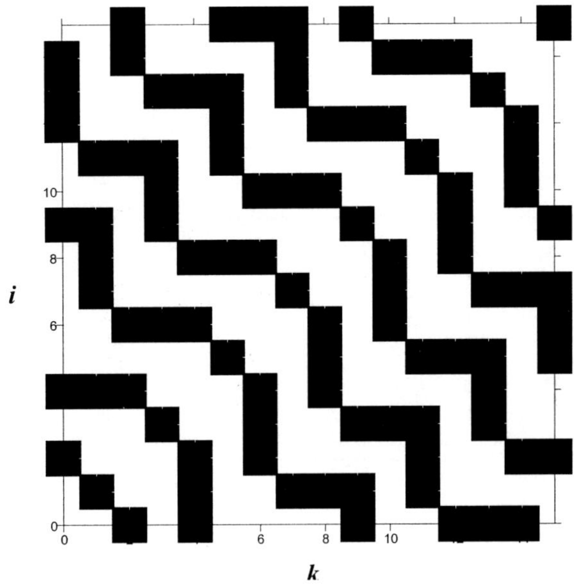

Figure 3.1 *One-dimensional, dual-coefficient basis function Wolfram Rule=15; N=16; Initial Configuration=0110110001011111; Boundary Configuration=Cyclic; CAT Type=2.*

Table 3.6 *Basis functions of the dual-coefficient transform generated with a 16-point Wolfram Rule 142 CA*

k→ / i↓	0	1	2	3	4	5	6	7	8	9	10	11	12	13	14	15
0	+	+	-	-	-	+	+	-	+	-	+	-	-	-	-	-
1	+	-	-	+	-	+	-	-	-	-	-	-	+	+	-	+
2	-	-	+	+	-	-	-	+	+	-	+	-	+	-	-	-
3	-	+	+	-	-	+	-	+	-	-	-	-	-	-	+	+
4	-	-	-	-	+	+	-	-	-	+	+	-	+	-	+	-
5	+	+	-	+	+	-	-	+	-	+	-	-	-	-	-	-
6	+	-	-	-	-	-	+	+	-	-	-	+	+	-	+	-
7	-	-	+	+	-	+	+	-	-	+	-	+	-	-	-	-
8	+	-	+	-	-	-	-	-	+	+	-	-	-	+	+	-
9	-	-	-	-	+	+	-	+	+	-	-	+	-	+	-	-
10	+	-	+	-	+	-	-	-	-	-	+	+	-	-	-	+
11	-	-	-	-	-	-	+	+	-	+	+	-	-	+	-	+
12	-	+	+	-	+	-	+	-	-	-	-	-	+	+	-	-
13	-	+	-	-	-	-	-	-	+	+	-	+	+	-	-	+
14	-	-	-	+	+	-	+	-	+	-	-	-	-	-	+	+
15	-	+	-	+	-	-	-	-	-	-	+	+	-	+	+	-

Two-Dimensional Bases

The most interesting in this family are the canonical forms derived from the evolving fields of one-dimensional automata. The relatively manageable alphabet base is a particularly attractive property of the canonical types. In Table 3.7 we show the gateway values for generating a select group of orthogonal dual-state, two-dimensional CA dual-coefficient basis functions.

Table 3.7 *Gateway Values for select two-dimensional dual-coefficient bases*

Wolfram Rule Number	Type	Initial Configuration	N
14	8	01001101	8
14	8	11010100	8
142	8	01101010	8
47	8	11110011	8
15	8	10010000	8
11	8	00111111	8
43	8	01111110	8
143	8	01011101	8
158	8	11011000	8
159	8	01011101	8
42	8	01110101	8
43	8	11010101	8
112	8	01011101	8
113	8	00101010	8
171	8	10101000	8
15	8	0110110001011111	16
142	*8*	*1101011010100101*	16

A graphical view of the basis functions generated using a set of the gateway values (Wolfram *Rule* 14; *Initial Configuration* 01001101) is shown in Figure 3.2.

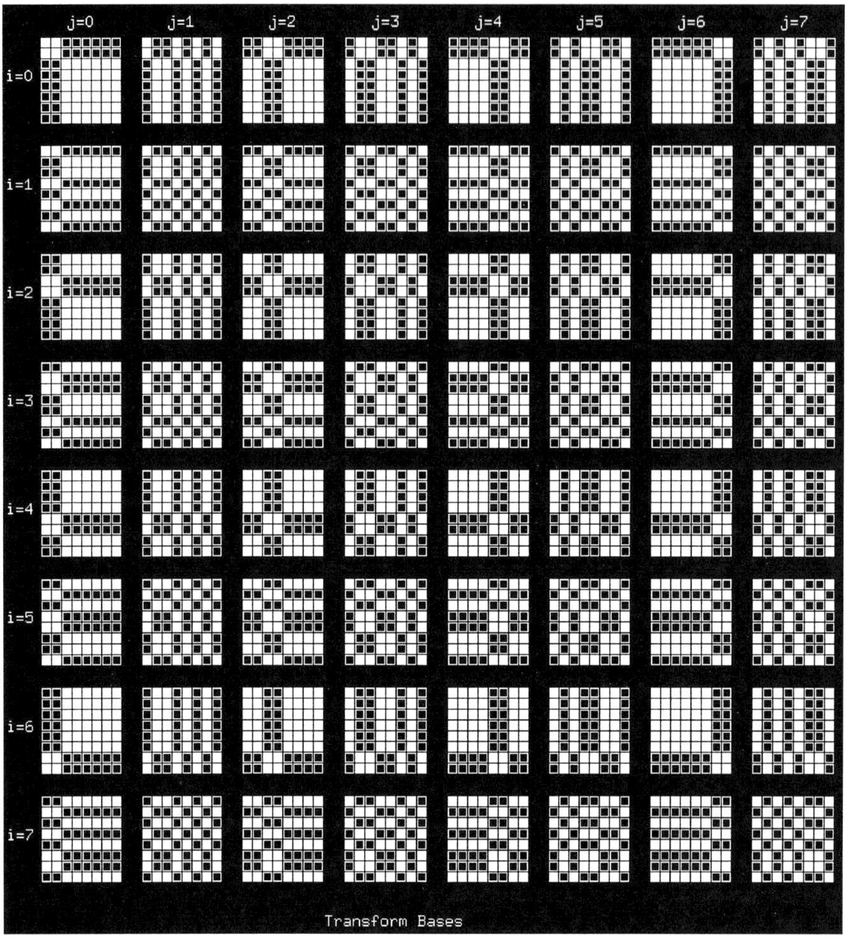

Figure 3.2 *Two-dimensional A_{ijkl} dual-coefficient basis functions. A_{00kl} is the block at the extreme upper left corner. The top row represents $0 \leq j < 8$; i=0. The left column is j=0; $0 \leq i < 8$. A_{ij00} is the upper left corner of each block. The white rectangular dots represent 1 (**addition**) while the black dots are -1 (**subtraction**).*

Fast CAT Transform

In general, the number of arithmetic operations involved in cellular automata transforms is of the order of N^2. However, the nature of the dual-coefficient CAT basis functions allows for fast transforms of the order cN where $c \leq log_2 N$. A sample C-code for a select fast dual-coefficient cellular automata transforms is shown in Appendix B.

Four-Site Neighborhood, Dual-Coefficient Bases

The evolution of the four-site neighborhood, dual-state cellular automaton is expressible in the form:

$$a_{it+1} = \begin{cases} F\left(a_{it}, a_{i+1t}, a_{i+2t}, a_{i+3t}\right) & \textbf{GroupI} \\ F\left(a_{i-1t}, a_{it}, a_{i+1t}, a_{i+2t}\right) & \textbf{GroupII} \\ F\left(a_{i-2t}, a_{i-1t}, a_{it}, a_{i+1t}\right) & \textbf{GroupIII} \\ F\left(a_{i-3t}, a_{i-2t}, a_{i-1t}, a_{it}\right) & \textbf{GroupIV} \end{cases} \tag{3.3}$$

where F is the Boolean function defining the rule.

There are 2^4=16 possible configurations for each neighborhood for dual-state, four-site neighborhood automaton. These are:

CONFIGURATION	KEY
1110	C_0
1100	C_1
1010	C_2
1000	C_3
0110	C_4
0100	C_5
0010	C_6
0000	C_7
1111	C_8
1101	C_9
1011	C_{10}
1001	C_{11}
0111	C_{12}
0101	C_{13}
0011	C_{14}
0001	C_{15}

in which C_n is the Boolean value generated by the rule given the n-th configuration. There are 2^{16} rules for the two-state/four-site CA.

A listing of the *rules* (written in their binary form) and *initial configuration* (with cyclic boundary conditions) for a small family of orthogonal Group I 4-site neighborhood, dual-state CA is shown in Table 3.8.

Table 3.8 *Gateway values for select two-dimensional, four-site dual-state, dual-coefficient transforms with cyclic boundary conditions*

Wolfram Rule Number	Type	Initial Configuration	N
0100010010111110	8	01111011	8
0101110010001011	8	11110001	8
0101011010011110	8	10000111	8
1101101101100111	8	11010000	8
0101011110011111	8	00011111	8
0101011110011011	8	11110111	8

3.2 Multi-Coefficient CA Basis Functions

By multi-coefficient basis functions, we mean filters whose values are not as simple as the (1,-1) presented above. We have found the Class II Scheme presented in Chapter 2 to be an excellent way of generating the multi-coefficient filters from the evolving field cellular automata. An additional degree of freedom is provided because of the multi-value nature. It is relatively easy to impose the additional constraints that may be required such as the degree of smoothness of the ensuing filters. The method through which we scale the filters so as to satisfy the additional conditions (*e.g.*, smoothness, orthogonality, overlapping) is algebraically involved and will not be elaborated upon here.

Non-Overlapping Filters

The following one-dimensional orthogonal, non-overlapping basis functions have been generated from a 16-cell, 32-state cellular automata. The filters are obtained using Type 1 Scheme II. The CA is evolved through eight time steps. The properties are summarized in Table 3.9.

Initial Configuration: 9 13 19 13 7 20 9 29 28 29 25 22 22 3 3 18
W-set: 0 13 27 19 26 25 17 5 14 1

Table 3.9 *Non-overlapping CAT filters*

k→ i ↓	0	1	2	3
0	0.8282762765884399	0.5110409855842590	0.1938057541847229	-0.1234294921159744
1	0.5476979017257690	-0.7263893485069275	-0.1903149634599686	0.3690064251422882
2	-0.1181457936763763	0.1970712691545487	0.5122883319854736	0.8275054097175598
3	-0.0051981918513775	0.4151608347892761	-0.8147270679473877	0.4047644436359406

Multi-dimensional, non-overlapping filters are easy to obtain by using canonical products of the orthogonal one-dimensional filters. Such products are not automatically derivable in the case of overlapping filters.

Overlapping CAT Filters

The following two-dimensional overlapping basis functions have been generated from a 16-cell, eight-state cellular automata using the Type 2 of Scheme II. The properties are summarized in Table 3.10.

Initial Configuration: 4 6 4 1 0 1 6 1 2 7 5 3 5 1 0 5
W-set: 6 7 5 3 4 4 7 0 1

The CA was evolved over eight time steps.

The following scaled transform coefficients are obtained from the states of the cellular automata evolved by using the above rule.

Table 3.10 *Forward 2D overlapping CAT filters*

i→ l+4k ↓	0	1	2	3
0	0.0000000000000000	0.0000000000000000	0.0000000000000000	0.0000000000000000
1	0.0000000000000000	0.0000000000000000	0.0000000000000000	0.0000000000000000
2	0.0000000000000000	0.0000000000000000	0.0000000000000000	0.0000000000000000
3	0.0000000000000000	0.0000000000000000	0.0000000000000000	0.0000000000000000
4	0.0000000000000000	0.0000000000000000	0.0000000000000000	0.0000000000000000
5	0.8333333730697632	-0.3726780116558075	-0.3726780116558075	0.1666666716337204
6	0.3333333432674408	0.7453560233116150	-0.1490712016820908	-0.3333333432674408
7	-0.1666666716337204	-0.3726780116558075	0.0745356008410454	0.1666666716337204
8	0.0000000000000000	0.0000000000000000	0.0000000000000000	0.0000000000000000
9	0.3333333432674408	-0.1490712016820908	0.7453560233116150	-0.3333333432674408
10	0.1333333253860474	0.2981424033641815	0.2981424033641815	0.6666666865348816
11	-0.0666666626930237	-0.1490712016820908	-0.1490712016820908	-0.3333333432674408
12	0.0000000000000000	0.0000000000000000	0.0000000000000000	0.0000000000000000
13	-0.1666666716337204	0.0745356008410454	-0.3726780116558075	0.1666666716337204
14	-0.0666666626930237	-0.1490712016820908	-0.1490712016820908	-0.3333333432674408
15	0.0333333313465118	0.0745356008410454	0.0745356008410454	0.1666666716337204

The inverse filters are obtained via a numerical inversion from the forward overlapping filters.

Table 3.11 *Inverse 2D overlapping CAT filters*

i→ l+4k ↓	0	1	2	3
0	0.2083333432674408	-0.0000000069538757	-0.0000000069538761	0.0000000000000001
1	0.4658475220203400	0.0000000067193628	0.0000000260078288	-0.0000000000000002
2	0.2083333581686020	0.4166667163372040	0.0000000149011612	0.0000000082784224
3	-0.0931694954633713	-0.1863389909267426	-0.0000000066640022	-0.0000000037022230
4	0.4658475220203400	0.0000000021659723	0.0000000029940725	-0.0000000000000000
5	1.0416667461395264	-0.0000000153978661	-0.0000000094374020	0.0000000000000001
6	0.4658474624156952	0.9316949248313904	-0.0000000066640018	-0.0000000037022230
7	-0.2083333134651184	-0.4166666269302368	0.0000000029802323	0.0000000016556845
8	0.2083333730697632	0.0000000079472855	0.4166667163372040	0.0000000082784224
9	0.4658474624156952	-0.0000000066916819	0.9316949248313904	-0.0000000037022232
10	0.2083333134651184	0.4166666567325592	0.4166666567325592	0.8333332538604736
11	-0.0931694731116295	-0.1863389462232590	-0.1863389760255814	-0.3726779520511627
12	-0.0931695029139519	-0.0000000066916819	-0.1863390058279038	-0.0000000037022232
13	-0.2083332985639572	0.0000000028560558	-0.4166666269302368	0.0000000016556845
14	-0.0931694880127907	-0.1863389760255814	-0.1863389760255814	-0.3726779520511627
15	0.0416666567325592	0.0833333283662796	0.0833333209156990	0.1666666567325592

3.3 *S*-Bases

The inherent similarity, which is an indication of redundancy, that exists amongst different parts of a given data sequence can be exploited in a compact encoding of the data by using Cellular Automata *S*-basis functions. In certain instances it is possible, starting from an arbitrary/random function, to achieve an accurate reconstruction of the encoded data via an iterative transformation. We refer to the first property as self-similarity. The second is self-generation. Self-similarity is a useful property for self-generation.

The concept of self-similarity is the basis of the fractal image compression method described by *Barnsley* [1993], *Barnsley & Hurd* [1993], and *Fisher* [1995]. The degree to which a given data is self-similar can be measured using an appropriate metric. Self-generation cannot always be guaranteed even when the magnitude of the error, in the self-similar transformation of the data, is small. Therefore, we resort to the concept of a *strong* or *weak* existence of self-generation. The self-generating property of a given data with respect to a given transformation is strong if it is always possible, starting from an initial arbitrary sequence, to recover the data by a repeated application of the *S*-transformation. Self-generation is weak when the recovery of the encoded data can only be achieved from a specified initial data set.

As far as cellular automata transforms are concerned, the chief attraction of a weak self-generating property is the ease with which convergence during decoding can be guaranteed by the CA selection process in the encoding phase of the processing. We select only CA gateway values that ensure convergence to the given data, starting from a specific initial set. The initial set may be formed by assigning the same constant value for all data points.

S-bases can be windowed or non-windowed. Computer experimentations show that non-windowed bases are best for carrying out self-generating transformation of data. The transform coefficients consist of scaling parameters that connect groups of data points with other groups within the same data sequence. *S*-transformation can result in a significant compression of the data.

One-Dimensional *S*-Transformation

Consider a data sequence f that has been grouped into M segments. In general, the segments may overlap. The overlapping of segments will yield a greater fidelity in encoding at the expense of an increased number of parameters and increased encoding time. In the following derivation we take the segments as non-overlapping. Let each consist of N data points. We transform the data, using the associated CA bases S, in the form:

$$f_{in} = \sum_{m=0}^{M-1} \lambda_{mn} \sum_{k=0}^{N-1} f_{km} S_{ik} + \varepsilon_{in} \qquad \begin{matrix} n = 0,1,2,\cdots M-1 \\ i = 0,1,2,\cdots N-1 \end{matrix} \tag{3.4}$$

in which the λ-values are scaling parameter, ε is the point-wise error in the representation, while f_{in} implies the value of f at the i-th data point of the n-th segment.

One immediate implication of equation (3.4) is obvious:

> The data value at a given point is a weighted sum of the entire data sequence. The weights consist of the CA basis functions and scaling parameters whose values must be determined.

The validity of equation (3.4) depends on the magnitude of the error ε involved in the transformation. A suitable measure of the overall error is the root-mean-squared (RMS) error:

$$E_s = \sqrt{\frac{1}{NM} \sum_{N=0}^{M-1} \sum_{i=0}^{N-1} \varepsilon_{in}^2}$$

Given a suitable *S*-basis, the λ-values can be determined by minimizing the sum of the squared errors:

$$\sum_{n=0}^{M-1}\sum_{i=0}^{N-1}\varepsilon_{in}^{2}=\sum_{n=0}^{M-1}\sum_{i=0}^{N-1}\left(f_{in}-\sum_{m=0}^{M-1}\lambda_{mn}\sum_{k=0}^{N-1}f_{km}S_{ik}\right)^{2} \tag{3.5}$$

which requires the solution of the following system of equations:

$$\sum_{n=0}^{M-1}\sum_{i=0}^{N-1}\sum_{m=0}^{M-1}p_{ij}p_{im}\lambda_{mn}=\sum_{n=0}^{M-1}\sum_{i=0}^{N-1}p_{ij}f_{in} \qquad j=0,1,2,\cdots,M-1$$

$$p_{ij}=\sum_{k=0}^{N-1}f_{kj}S_{ik} \tag{3.6}$$

The above representation assumes that the data belonging to a given segment can be generated from a transformation involving the entire data. For data compression purposes, we want to find CA bases that will maximize the number of insignificant or zero λ-values. An ideal case is when:

$$\lambda_{mn}=\delta_{mn_0}\lambda_{m_0 n}$$

$$\delta_{mn_0}=\begin{cases}1 & m=m_0\\ 0 & m\neq m_0\end{cases} \tag{3.7}$$

in which $m_0=m_0(n)$. In that case, equation (3.4) degenerates into:

$$f_{in}=\lambda_{m_0 n}\sum_{k=0}^{N-1}f_{km}S_{ik}+\varepsilon_{in} \qquad \begin{array}{l} n=0,1,2,\cdots M-1\\ i=0,1,2,\cdots N-1\end{array} \tag{3.8}$$

Thus, the data belonging to segment n is generated from scaled/weighted values of the data belonging to just one other segment, m_0. Another implication of *S*-transformation is obvious by comparing equation (3.8) with the "conventional" CA transform equation:

$$f_i=\sum_k c_k A_{ik}$$

The coefficients in S-transformation are scaled values of data belonging to select parts of the data sequence. As a matter of fact, the S-bases are generated in the same way as the A bases discussed in earlier chapters.

Using equation (3.7) in (3.6), we obtain:

$$\lambda_{m_0 n} = \frac{\sum_{n=0}^{M-1} \sum_{i=0}^{N-1} p_{im_0} f_{in}}{\sum_{i=0}^{N-1} p_{im_0}^2} \tag{3.9}$$

and the RMS error for the segment is

$$E_s = \sqrt{\frac{1}{N} \sum_{i=0}^{N-1} \left(f_{in} - \lambda_{m_0 n} \sum_{k=0}^{N-1} f_{km_0} S_{ik} \right)^2} \tag{3.10}$$

The encoding steps are:

1. Randomly select a set of CA gateway values:
   - Rule number
   - Initial configuration
   - Boundary configuration
   - Base type
   - Window size, N_w, if the bases are windowed
   - The length scale, L_w, if required
2. For each segment $n = 0,1,2,\ldots M-1$, find the segment $m_0 = m_0(n)$ and the scaling parameter $\lambda_{m_0 n}$ for which the transformation error E_s is minimum.

3. Starting with an initial data f_0, reconstruct the data by a repeated application of equation (3.8). Record the reconstruction error:

$$E_f = \sqrt{\left\{ \frac{1}{NM} \sum_{n=0}^{M-1} \sum_{i=0}^{N-1} \left(f_{in} - f_{in}^{q} \right)^2 \right\}} \tag{3.11}$$

in which q is the number of times the transformation is applied.

4. Note the errors, E_s and E_f, and associated parameters λ and m_0 for all segments.

5. Repeat the above operations for a sufficiently large number of trial gateway values.

6. Select the transformation with the best overall E_s and E_f values.

The encoding requires the storage or transmission of the gateway values used to generate the basis functions S, and the optimal $m_0 = m_0(n)$ and λ-values.

The reason for reconstructing the data during the encoding process is to ensure that the repeated application of the S-transformation will actually converge to the original data. In the reconstruction of the data f, we start with a trial function f^0, and obtain:

$$f_{in}^{1} = \lambda_{m_0 n} \sum_{k=0}^{N-1} f_{km_0}^{0} S_{ik} \qquad \begin{array}{l} n = 0,1,2,\ldots M-1 \\ i = 0,1,2,\ldots N-1 \end{array}$$

from which an iterative calculation of f^2, f^3, $\ldots f^{\infty}$ can be achieved. At the I-th iteration level:

$$f_{in}^{I} = \lambda_{m_0 n} \sum_{k=0}^{N-1} f_{km_0}^{I-1} S_{ik} \qquad \begin{array}{l} n = 0,1,2,\ldots M-1 \\ i = 0,1,2,\ldots N-1 \end{array} \tag{3.12}$$

Let $d_I = \left\| f^I - f^{I-1} \right\|_{in}$ be the distance between the values of the generated data at iteration levels *I* and *I*-1, respectively.

The quality of the *S*-transformation is good if the following conditions are satisfied:

1. The self-similarity error, E_s, is minimal.

2. The reconstruction process is uniformly convergent. That is:

$$d_I < d_{I-1} < d_{I-2} < \cdots d_3 < d_2 < d_1 \quad \text{for any } I$$

which also implies that after a large number of repeated applications of equation (3.12):

$$d_\infty \to 0$$

3.The reconstruction error, E_f, is minimal.

With a strong self-generating property, the above conditions will be satisfied regardless of the choice of f^0. The *S*-transformation is weak when the convergence of the solution is bases on a particular selection of f^0.

Orthogonal Decomposition of the Errors

The error ε in equation (3.4) or (3.8) can be represented by orthogonal CA bases:

$$\varepsilon_{in} = \sum_{k=0}^{N=1} c_{kn} A_{ik} \tag{3.13}$$

$$c_{kn} = \frac{\sum_{i=0}^{N-1} \varepsilon_{in} A_{ik}}{\sum_{i=0}^{N-1} A_{ik}^2} \tag{3.14}$$

where the bases A, not necessarily the same as the bases used in the *S*-transformation, are derived from orthogonal *windowed* or *non-windowed* basis functions.

The goal in data compression applications is that the errors involved in the *S*-transformation will be so small that the majority of the coefficients c_{kn} will be negligible.

By including the orthogonal decomposition of the errors, the complete *hybrid* transformation is obtained by combining equations (3.8) and (3.14) as:

$$f_{in} = \lambda_{m_0 n} \sum_{k=0}^{N-1} f_{km} S_{ik} + \sum_{k=0}^{N-1} c_{kn} A_{ik} \qquad \begin{array}{l} n = 0,1,2,\ldots M-1 \\ i = 0,1,2,\ldots N-1 \end{array} \tag{3.15}$$

Observe that when there is no *S*-transformation, $\lambda=0$, equation (3.15) degenerates into the conventional CA transform statement. With the above hybrid formulation, the error can be as small as required depending on the price we are willing to pay for an increased number of transform coefficients.

Two-Dimensional S-Transformation

The extension of the above development to two and more dimensions is not difficult. Given a two-dimensional data array f, we subdivide these into M blocks each of size N x N. The quantity f_{ijn} is the data value at the point (i,j) belonging to the n-th element. The *S*-transformation will be in the form:

$$f_{ijn} = \lambda_{m_0 n} \sum_{k=0}^{N-1} f_{klm_0} S_{ijkl} + \varepsilon_{ijn} \qquad \begin{array}{l} n = 0,1,2,\cdots M-1 \\ i,j = 0,1,2,\cdots N-1 \end{array} \tag{3.16}$$

where the one-parameter-per-segment approach has been used. The bases S_{ijkl} are formed in the same way those outlined in Chapter 2. Orthogonality is not necessarily a nice feature in *S*-transformation.

Rather, the ability of the transformation to converge (small E_f) into a close copy of the original data is the most desired property.

Hybrid Two-Dimensional Transformation

As in the one-dimensional case, the hybrid formulation is obtained when the error ε is represented by orthogonal two-dimensional bases A_{ijkl} as:

$$e_{ijn} = \sum_{k=0}^{N-1} \sum_{l=0}^{N-1} c_{kln} A_{ijkl} \tag{3.17}$$

$$c_{kln} = \frac{\sum_{i=0}^{N-1} \sum_{j=0}^{N-1} \varepsilon_{ijn} A_{ijkl}}{\sum_{i=0}^{N-1} \sum_{j=0}^{N-1} A_{ijkl}^2} \tag{3.18}$$

Hence, the complete hybrid transformation is:

$$f_{ijn} = \lambda_{m_0 n} \sum_{k=0}^{N-1} f_{klm_0} S_{ijkl} + \sum_{k=0}^{N-1} \sum_{l=0}^{N-1} c_{kln} A_{ijkl} \qquad \begin{array}{l} n = 0,1,2,\ldots M-1 \\ i = 0,1,2,\ldots N-1 \end{array} \tag{3.19}$$

Multi-Resolution Analysis

Self-generation provides an excellent tool for carrying out multi-resolution analysis. Consider a data sequence f_i defined on a regular lattice of cells $i = 0,1,2,\ldots N-1$. Suppose we want to estimate the value of the function at the mid-node points:

$$j = i + \frac{1}{2} \qquad i = 0,1,2,\ldots N-1$$

A most natural approach is to assume:

$$f_i = \frac{f_i + f_{i+1}}{2} \qquad i = 0,1,2,\ldots \tag{3.20}$$

using a linear interpolation of f between the N nodes at which the value of the function is originally supplied. The S-transformation provides another way by which the required interpolation can be achieved. If the above segment is the n-th segment of a large M-segment data sequence, then with S-transformation there is a linkage between the data sequences belonging to segment n and those belonging to another segment m_0 in the form of equation (3.8):

$$f_{in} = \lambda_{m_0 n} \sum_{k=0}^{N-1} f_{km_0} S_{ik} \qquad \begin{array}{l} n = 0,1,2,\ldots M-1 \\ i = 0,1,2,\ldots N-1 \end{array} \tag{3.21}$$

Self-similarity is a property of both segments n and m_0, and should hold regardless of the number of points selected on each segment. Therefore, by adding the mid-nodes into the original lattice (effectively doubling the number of cells) we can write equation (3.21) for the new lattice as:

$$f_{inR} = 2^{-R} \lambda_{m_0 n} \sum_{k=0}^{2^R(N-1)} f_{km_0 R} S_{ik} \qquad \begin{array}{l} n = 0,1,2,\ldots M-1 \\ i = 0,1,2,\ldots 2^R(N-1) \end{array} \tag{3.22}$$

where R is the resolution. The original lattice corresponds to R=0. Equation (3.21) is, in essence, written for the coarsest resolution R=0. Compatibility demands that:

$$f_{(2i)nR} = f_{in(R-1)} \qquad i = 0,1,2,\ldots$$

since those points are physically the same at the different resolutions.

Equation (3.22) is valid for any resolution R=0,1,2,3,... Starting with, say, equation (3.19) as an initial estimate, equation (3.22) can be applied repeatedly to generate new mid-node values. The iterative operation is halted when a stipulated convergence criterion is achieved. Having obtained the values of f at the mid-node points, the analysis can be carried out to the quarter-points (R=2) of each segment. The process can theoretically be carried out *ad infinitum*.

In order to implement the analysis to a fine resolution at which $R = R_f > 0$, the basis S must be defined at the finest lattice. That way we are sure that the S-values will be available at all lower resolution. Otherwise, we will have to estimate fine scale S-values from coarse-scale quantities. Although quite acceptable results have been obtained using coarse-fine scale approximations of the S-bases, the best is to generate the CA basis functions from the largest number of cells, $2^{R_f}(N-1)+1$, belonging to the stipulated finest level.

The above procedure is a valuable tool in *zooming* on select parts of a function or a digitized image. We are able to get as much information as possible from any selected region of the function/image. Granted, the new information has come from a mathematical deduction, *albeit* using self-similarity. However, the technique exploits the self-generating character of the data sequence. The approach allows us to expand image data with a relatively few number of encoding parameters, *i.e.*, λ_{m0n} and m_0. Huge "compression ratios" can therefore be claimed by calculating the number of bits that will otherwise be required to encode the expanded region of the image. In Chapter 4 we present another zooming method that utilizes sub-band CAT bases.

Part II

Applications of Cellular Automata Transforms

In the following chapters we showcase a few practical applications of Cellular Automata Transforms. These applications include multimedia compression, data encryption, and solution of differential and integral equations.

Chapter 4

Multimedia Compression

4.1 Introduction

Data compression schemes seek to remove or minimize the inherent redundancy within a data string. In general, efforts at compressing any data invariably comes at a price. As depicted in Figure 4.1, the cost pertains to three interrelated factors:

1. Compression ratio. The data storage space and/or data transmission time vary inversely with the compression.

2. Computational time. How many mathematical operations are involved? Adaptive codes, for example, will generally provide better compression ratios and/or improved fidelity. The cost is the time (or computational effort) required to attain the improvement.

3. Encoding/Decoding fidelity. Is there any loss of information during the encoding/decoding phases?

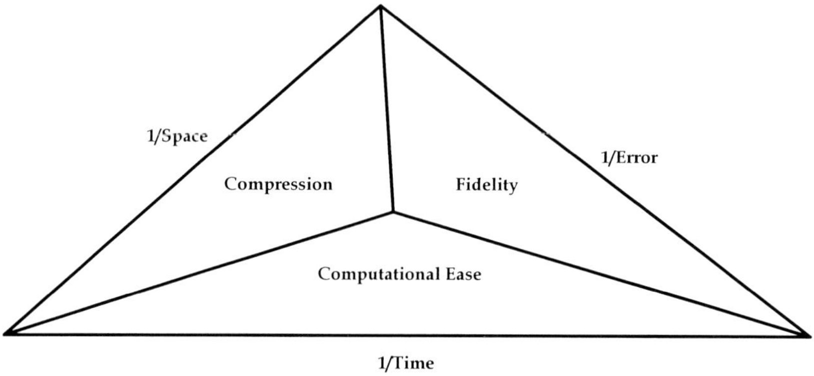

Figure 4.1 *Interrelated factors that influence the compression process*

Obviously, in situations where the compression process requires a perfect reconstruction of the original data (lossless encoding), the error size must be zero. In that case Figure 4.1 degenerates into a dual-factor relationship in which more compression is typically achieved at the expense of computational ease. A classic example is the family of adaptive encoders.

Data that is perceived (*e.g.*, photographs, audio, and video) can often be compressed with some degree of loss in the reconstructed data. Greater compression is achieved at the expense of signal fidelity. In this case a successful encoding strategy will produce an error profile that cannot be perceived by the human eye (digital images and video) or ear (digital audio). Perceptual coding becomes a key and integral part of the encoding process.

In using Cellular Automata Transforms to compress data, the redundancy is identified by transforming the data into the CA space. The principal strength of CAT-based compression is the large number of transform bases available. We make use of CA bases that maximize the number of transform coefficients with insignificant magnitudes. We may also desire a transform that always provides a predictable global pattern in the coefficients. This predictability can be taken advantage of in optimal bit assignment for the coefficients.

Therefore, CAT permits the selection of basis functions that can be adapted to the peculiarities of the data. A principal strength of CA encoding is the parallel and integer-based character of the computational process involved in evolving states of the cellular automata. This can translate into an enormous computational speed in a well-designed, CAT-based encoder.

Apart from the compression of data, CAT also provides excellent tools for performing numerous data processing chores, such as digital image processing (*e.g.*, image segmentation, edge detection, image enhancement) and data encryption.

In this chapter we present the fundamentals of lossy data compression using Cellular Automata Transforms. We will outline the various strategies we have developed in using CAT to compress digital images, audio and video.

Approaches in CAT Data Compression

Given a data sequence f_i, all the CA transform techniques seek to represent the data in the form:

$$f_i = \sum_k c_k A_{ik} \tag{4.1}$$

in which c are transform coefficients, while A are the transform bases. The basic strategy for compressing data using CA is:

- Start with a set of CA gateway keys that produce basis functions A and its inverse B.

- Calculate the transform coefficients.

- For lossy encoding, quantize the coefficients. In this approach the search is for CA bases that will maximize the number of negligible transform coefficients. The energy of the transform will be concentrated on the few retained coefficients. Ideally there will be a different set of CA gateway values for different parts of a data file. There is a threshold point at which the overhead involved in keeping track of different gateway values far exceeds the benefit gained in greater compression or encoding fidelity. In general, it is sufficient to "initialize" the encoding by searching for the one set of gateway keys with nice overall properties: *e.g.*, orthogonality, maximal number of negligible transform coefficients and predictable distribution of coefficients for optimal bit assignment. This approach is the one we will normally follow in most of CA data compression schemes. The encoding parameters include the gateway keys and the CA transform coefficients.

4.2 Encoding Strategy

The existence of trillions of transform bases and CA gateway keys mandates different strategies for different data encoding tasks. For each task, a decision must be made as to how many different CA bases will be used. An evaluation must also be made of the cost associated with each decision in terms of computational cost, encoding/decoding time and fidelity. The three major schemes are the following:

1. **Single CA Base for Entire Data File:** Each data block is encoded and decoded using the same CA basis functions. The advantage is that the gateway keys for the CA can be embedded in the CA coder/decoder, not in the compressed file. This is the best approach for tasks where speed is of the essence. The cost is the inability to fully exploit the main strength of CA adaptability. This is a symmetric process, since the encoding and decoding will take approximately the same amount of time. The single base per file circumvents the need to design different quantization strategies for different parts of the data. The same CA basis can even be used for several data files just as, for instance, the discrete cosine transform is utilized for a variety of image compression tasks.

2. **Single CA Base for Select Regions of Data File:** Entire groups of data blocks are encoded with the same CA basis functions. While there is some sacrifice in time as a result of generating different CA bases for different regions, the number of gateway keys can be kept small; at least much smaller than the number of data blocks. This approach takes some advantage of the adaptive strength of CA encoding. The compression ratio and encoding fidelity will be much better than in (1). The encoding time will be slightly more than the decoding time. The degree of asymmetry will be dictated by the number of different basis functions used.

3. **One CA Base per Data Block:** This approach is excellent in tasks where massive compression is the primary goal. The time it takes to search for the best CA basis functions for each data block makes this most suitable for *off-line* asymmetric encoding tasks. An example is the publication of data on CD-ROMs where the encoded file will be read many times, but the compression is done once. We can then expend as much computational resources and time as we can get searching for the best CA basis functions for each block of the data file. The multiple base approach fully exploits the adaptive strength of CA Transforms.

Optimal CA Keys Selection Criteria

In the following presentation, our attention is focussed on Strategy 1, where the same set of CA keys is used to encode the entire file. Once an optimal set of CA keys has been selected for a test data, the same set can be used routinely to encode other data.The entropy of the transform coefficients is:

$$\boldsymbol{E} = -\sum_{i} \phi_i \log_2(\phi_i) \tag{4.2}$$

where ϕ_i is the probability that a transform coefficient is of magnitude i. To compute ϕ_i, we calculate the number of times the transform coefficients attain the value i (*i.e.*, the frequency of i). The resulting value is divided by the sum of all frequencies to obtain the pertinent probability. The goal is to find CA keys that will result in the minimization of the entropy of the transform coefficients.

The quantization strategy is a function of how the data will be perceived. For digital images and video, low frequencies are given a higher priority than high frequencies because of the way the human eye perceives visual information. For digital audio, both low and high frequencies are important and the coefficient decimation will be guided by a psycho-acoustics profile.

4.3 Digital Image Compression

In the following we assume the CAT filters are of the sub-band type. The analysis holds for both overlapping and non-overlapping filters. Let $w=2^n$ be the width (the number of pixels) of the image while $h=2^m$ is the height, where m,n are integers. Dimensions that are not integral powers of two are handled by the usual zero-padding method. The transform coefficients c_{kl} fall into four distinct classes (see Figure 4.2a): Those at even k and l locations (Group I) represent the "low frequency" components. These are sorted to form a new image of size $2^{(n-1)}2^{(m-1)}$ (at a lower resolution). The rest (Group II: k even, l odd; Group III: k odd, l even; Group IV: k odd, l odd) of the coefficients are "high frequency" components.

The low frequency, Group I components can be further transformed. The ensuing transform coefficients are again subdivided into four groups (Figure 4.2b). Those in Groups II, III, and IV are stored while Group I is further CAT-decomposed and sorted into another 4 groups. For an image whose size is an integral power of two, the hierarchical transformation can continue until Group I contains only one-quarter of the filter size. In general, the sub-band coding will be limited to n_R levels. Figure 4.2a represents the transform data at the finest resolution. The last transformation, at the n_R-th level, is the coarsest resolution.

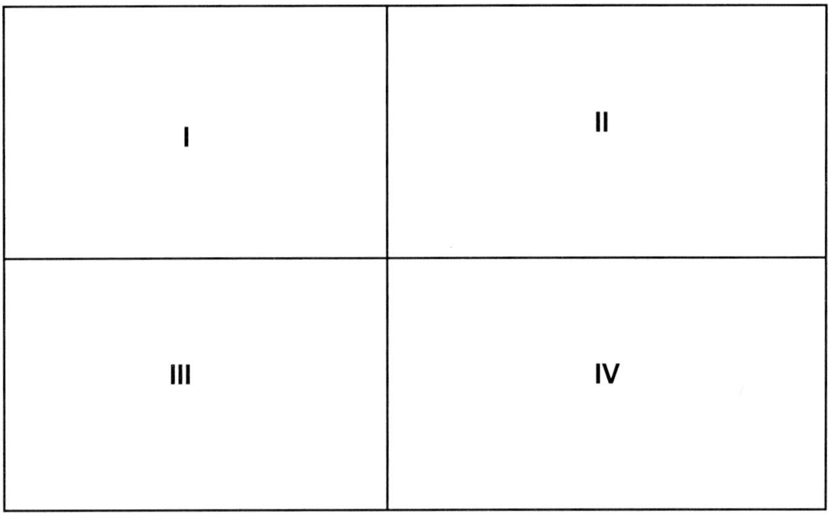

Figure 4.2a *Decomposition of CAT coefficients into four bands. Group I is equivalent to the "low frequency" components. Group II, III, and IV are the "high frequency" components.*

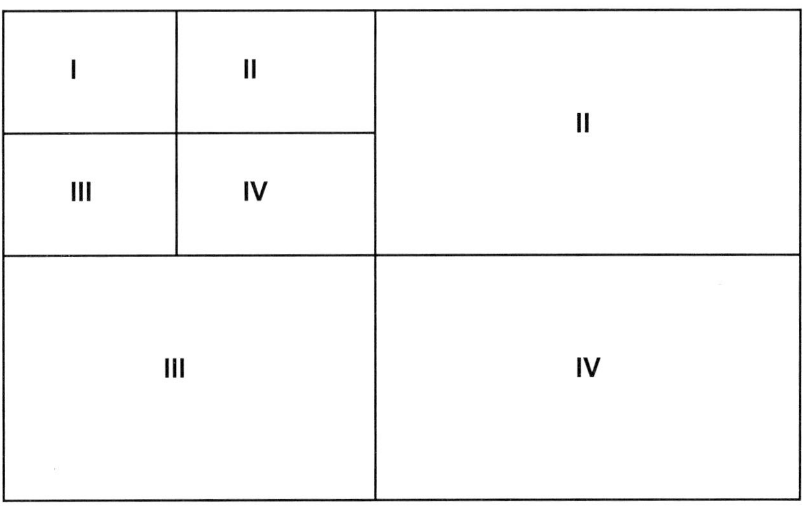

Figure 4.2b *The image formed by the Group I components in Figure 4.2a is further CAT decomposed and sorted into another four groups at the lower resolution.*

Edge Detection

By throwing away Group I coefficients ("low frequencies") and retaining only those in Groups II, III, and IV ("high frequencies"), we have a robust means for edge detection. Figures 4.3a,b show the use of dual-coefficient CAT filters in detecting the edge of a barn. Edge detection is a critical process in many applications including pattern/target recognition in biometrics and defense analysis.

Figure 4.3a *Original image of a barn*

Figure 4.3b *Image reproduced by using only CAT coefficients in Groups II, III, and IV at the highest resolution*

Zooming

The Group I low frequency coefficients provide the tool for zooming up or down on an image. Group I coefficients (with proper normalization) form the zoomed down image. So with $n_R = 1$ the forward transform produces Group I coefficients whose size is one-quarter (Figure 4.4b) of the original image (Figure 4.4a). The zooming process can continue by further transforming the reduced image and using the new set of Group I coefficients.

To zoom up on an image, the original image is assumed to be the Group I transform coefficients of the new larger image to be formed. The coefficients in Groups II, III and IV are set to zero. An inverse CA transform is carried out to recover a new image that is four times the size of the original (Figure 4.4c). The process can be repeated to produce an image that is 16 times the original.

Figure 4.4a *Original 319x215 Tiger*

Figure 4.4b *Zoomed down 159x107 Tiger*

Figure 4.4c *Zoomed up 638x430 Tiger*

Compression Scheme

The nature of the transform coefficients derived from a sub-band CAT coder makes it possible to impose objective conditions based on either: 1) a target compression ratio; or 2) a target error bound. The beauty of an orthonormal transform is that the error in the reconstructed data is equal to the maximum discrepancy in the transform coefficients. The encoding philosophy for a sub-band coder is intricately tied to the cascade of coefficient Groups I, II, III, and IV shown in Figures 4.2a and b. The coding scheme is hierarchical. Bands at the coarsest levels typically contain coefficients with the largest magnitudes. Hence, the coding scheme gives the highest priority to bands with the largest coefficient magnitudes.

All the coding schemes make use of a three-symbol alphabet system:

- 0 (***YES***)
- 1 (***NO*** or ***POSV***)
- 2 (***NEGV***).

If a target compression ratio C_R is desired, the steps involved in the scheme are the following:

1. Calculate TargetSize = C_R . OriginalFileSize.
2. Determine T_{max}=magnitude of coefficient with the largest value throughout all the bands.
3. Set $Threshold = 2^n > T_{max}$, where n is an integer.
4. Output n. This number is required by the decoder.
5. Set OutputSize = 0.
6. Perform steps i, ii, and iii while $OutputSize<TargetSize$.

i. For each of the sets of data belonging to Groups I, II, III, and IV, march from the coarsest sub-band to the finest. Determine T_b = maximum coefficient in each sub-band.

ii. If T_b<*Threshold*, encode ***YES*** and move onto the next sub-band. Otherwise, encode ***NO*** and proceed to check each coefficient in the sub-band:

a) If the coefficient value is less than *Threshold*, encode ***YES***.

b) Otherwise, encode ***POSV*** if coefficient is positive or ***NEGV*** if it is negative.

c) Decrease the magnitude of the coefficient by *Threshold*.

iii. Set *Threshold* to *Threshold/2*. Return to step (i) if *OutputSize*<*TargetSize*.

If a target error E_{max} is the goal, the steps involved in the scheme are the following:

1. Determine T_{max}=magnitude of coefficient with the largest value throughout all the bands.

2. Set $Threshold = 2^n > T_{max}$, where *n* is an integer.

3. Output *n*. This number is required by the decoder.

4. Perform steps i, ii, and iii while *Threshold* > E_{max}.

i. For each of the sets of data belonging to Groups I, II, III, and IV, march from the coarsest sub-band to the finest. Determine T_b=maximum coefficient in each sub-band.

ii. If T_b < *Threshold*, encode ***YES*** and move onto the next sub-band. Otherwise, encode ***NO*** and proceed to check each coefficient in the sub-band.

a) If the coefficient value is less than *Threshold,* encode ***YES.***

b) Otherwise, encode ***POSV*** if coefficient is positive or ***NEGV*** if it is negative.

c) Decrease the magnitude of the coefficient by *Threshold.*

iii. Set *Threshold* to *Threshold/2*. Return to step (i) if *Threshold* > E_{max}.

Symbol Packing Strategy and Entropy Coding

As the symbols ***YES, NO, POSV, NEGV*** are written, they are packed into a byte derived from a five-letter base-3 word. The maximum value of the byte is 242, which is equivalent to a string of five ***NEGV***. The above encoding schemes tend to produce long runs of zeros. The ensuing bytes can be entropy encoded using an Arithmetic Code or any of the Dictionary based methods. Otherwise, the packed bytes can be run-length coded and then the ensuing data is further entropy encoded using a special 16-bit word Huffman Code. The examples shown below utilize the latter approach.

Color Images

In color images, the data *f* is a vector of three components representing the primary colors RED (R), GREEN (G), and BLUE (B). Each of the colors can have any value between 0 and $2^b - 1$, where *b* is the number of bits per pixel. Each color component is treated the same way a grayscale data is processed. It is most convenient to work with the YIQ model, the standard for color television transmission. The Y-component stands for the *luminance* of the display, while the I- and Q-components denote *chrominance*. The luminance is derived from the RGB model using:

$$Y = 0.299R + 0.587G + 0.114B \tag{4.3}$$

The chrominance components are computed from:

$$I = 0.596R - 0.275G - 0.321B \tag{4.4}$$

$$Q = 0.212R - 0.523G + 0.311B \tag{4.5}$$

The advantage of the YIQ model is the freedom to encode the components using different degrees of fidelity. The luminance represents the magnitude of light being deciphered by the human eye. The I and Q components represent the color information. When the attainment of large compression ratios is a major goal, the chrominance components can be encoded with a much lower degree of fidelity than the luminance portion.

Compression Results

The Original 512x512 color Lena image (Figure 4.5a) has been selected to showcase the CAT image compression approach. The CAT filters used are those shown in Tables 3.10 and 3.11. The compressed files are shown to one-quarter scale in Figures 4.5b-h. The chief strength of CAT compression is the ability to maintain relatively smooth, non-pixelized images at very low bit rates (Figures 4.5e-h). Figure 4.6b shows a 45:1 compression of the Tiger.

Figure 4:5a *Original Lena, 786,486 bytes (1:1)*

Figure 4:5b *Compressed Lena, 86,802bytes (9:1)*

Figure 4:5c *Compressed Lena, 32,805 bytes (24:1)*

Figure 4:5d *Compressed Lena, 17,593 bytes (49:1)*

Figure 4:5e *Compressed Lena, 8,064 bytes (98:1)*

Figure 4:5f *Compressed Lena, 4,151 bytes (190:1)*

Figure 4:5g *Compressed Lena, 1,961 bytes (401:1)*

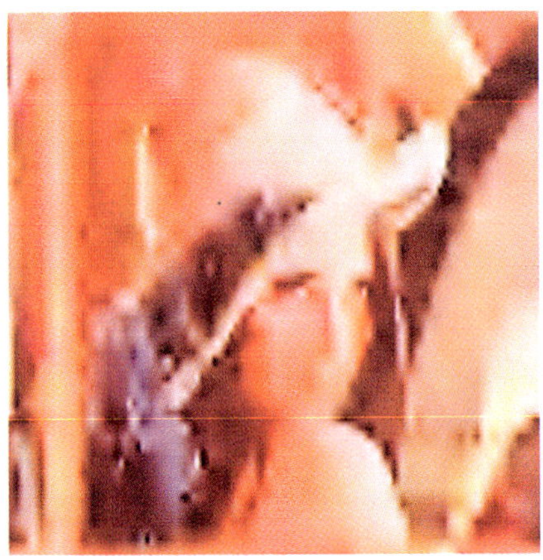

Figure 4:5h *Compressed Lena, 945 bytes (832:1)*

Figure 4.6a *Original Tiger 921,654 bytes (1:1)*

Figure 4.6b *Compressed Tiger 20,320bytes (45:1)*

4.4 Audio Compression

While an image coder must put a greater priority on low frequencies than on high frequencies, the audio coder has to deal with the complexity of the human audio perception system. There is the issue of the minimum threshold of hearing. When the strength of a given frequency falls below the threshold of hearing, that frequency can be removed without an adverse effect on the decoded sound. The importance of a specific audio frequency in a signal depends on the characteristics of the neighboring frequencies. Louder tones may drown softer tones in a phenomenom known as amplitude masking. Furthermore, the human ear is most sensitive to frequencies within certain ranges. *Pohlman* [1995] has a detailed presentation of the fundamental phenomena that control human hearing. An efficient audio compression scheme must take advantage of the peculiarities in human hearing.

The volume of data required to encode raw audio data is large. Consider a stereo audio music sampled at 44100 samples per second and with a maximum of 16 bits used to encode each sample per channel. A one-hour recording of a raw digital stereo music with that fidelity will occupy over 600Mb of storage space. To transmit such an audio file over 56kilobits per second communications channel (e.g., the rate supported by most POTS through modems), will take over 24 hours.

As far as CA-generated basis functions are concerned the non-overlapping filters tend to produce higher fidelity compressed audio signals than the overlapping filters. While audio data can be compressed using orthogonal or non-orthogonal non-sub-band CA basis functions, we will first showcase the use of orthogonal sub-band CA bases.

Sub-band CAT Audio Compression

The transform coefficients are grouped into low and high frequencies. The coder uses a sub-band thresholding method akin to the error-constrained approach previously described for digital images. Let T_e be the threshold at which the coding terminates for each sub-band. Then the audio coding scheme follows these steps:

1. Determine T_n, the maximum coefficient in the n-th sub-band $(n = 0,1,2,\ldots n_R - 1)$ where n_R is the number of sub-bands.

2. Perform Steps 3-5 for all the sub-bands for which $T_n > T_e$.

3. For each sub-band, set *Threshold* = $2^m > T_n$, where m is an integer.

4. Output m. This number is required by the decoder.

5. Perform steps i, ii, and iii while *Threshold* > T_e

    i. For each of the sets of data belonging to low and high frequency, march from the coarsest sub-band to the finest. Determine T_b = maximum residual coefficient in each sub-band.

    ii. If T_b < *Threshold,* encode ***YES*** and move onto the next sub-band. Otherwise, encode ***NO*** and proceed to check each coefficient in the sub-band:

    a) If the coefficient value is less than *Threshold* encode *YES*.

    b) Otherwise, encode ***POSV*** if coefficient is positive or ***NEGV*** if it is negative.

    c) Decrease the magnitude of the coefficient by *Threshold.* This results in a new residual coefficient.

iii. Set *Threshold* to *Threshold/2*. Return to step (i) if *Threshold* > T_e.

The decoding steps for this type of an embedded scheme are easy to implement. Decoding generally follows the natural order of the encoding process:

- Read n. Calculate *Threshold*=2^n. Calculate TargetSize=C_R.OriginalFileSize
- Set InputSize=0. Initialize all transform coefficients to zero.
- Perform steps i, ii, iii while InputSize < TargetSize

  i. March through the sub-bands from the coarsest to the finest.
  ii. Read ***CODE***. If ***CODE=YES,*** move onto the next sub-band. Otherwise proceed to decode each coefficient in the sub-band:
     a) Read ***CODE***
     b) If ***CODE=POSV,*** add *Threshold* to the coefficient
     c) If ***CODE=NEGV,*** subtract *Threshold* from the coefficient
  iii. Set *Threshold* to *Threshold/2*. Return to step (i) if InputSize < TargetSize.

Variants of the embedded sub-band scheme are used later in describing CAT compression of audio and video data. When appropriate we will only outline the encoding process from which the reader can infer the associated decoding steps.

The rate of decrement of the threshold can be made a function of the band, instead of the constant 50% used above. The termination threshold, T_e, is derived from psycho-acoustics models developed specifically for CAT-based audio filters. The model calculates the termination threshold as:

$$T_e = 2^{\frac{1}{n_R}\sum_{n=0}^{n_r-1}\omega_n \log_2(T_n)-Q} \tag{4.6}$$

where Q is an audio-fidelity parameter and ω_n $(n=0,1,2,\ldots n_R-1)$ are weights whose distribution defines the acoustic importance of each sub-band. The simplest model is obtained when the bands are given the same weight by setting $\omega_n=1$ for all sub-bands. Large values of Q correspond to higher audio quality but reduced compression. The termination threshold is a measure of the error introduced in the coding process. In the limiting case when $T_e \rightarrow 0$ (or $Q \rightarrow \infty$) we have a near lossless reconstruction of the audio data.

The non-overlapping, orthogonal, sub-band CAT filters shown in Table 4.1 have been evolved specifically for compressing audio data.

Table 4.1 *Non-overlapping CAT filters*

k→ i ↓	0	1	2	3
0	-0.8275159001350403	-0.5122717618942261	0.1970276087522507	0.1182165592908859
1	-0.2851759195327759	0.7287828922271729	0.6020380258560181	0.1584310680627823
2	0.1233587935566902	-0.1938495337963104	-0.5110578536987305	-0.8282661437988281
3	-0.4676266610622406	0.4109446406364441	0.5809907317161560	-0.5243086814880371

Table 4.2 shows the summary of the CAT compression of the first 8Mb of a "soft rock" music segment. The test section is a 16-bit, 44.1 KHz stereo music and it is divided into 463 segments ranging in length from 256 samples to 131,072 samples. The segments are formed with the objective of grouping samples of the same order of signal amplitude together.

Table 4.2 *Fidelity/compression/threshold profile*

Fidelity Parameter Q	Compression Ratio	Avg. Termination Threshold	Max. Termination Threshold
2	98.4	2208	8192
3	45.1	1104	4096
4	22.4	552	2048
5	12.1	276	1024
6	7.3	138	512
7	4.8	69	256
8	3.4	35	128

Table 4.3 shows the influence of n_R on the compression of the same music segment with Q=5.

Table 4.3 *Effect of* n_R *on compressed file size*

Number of Sub-bands, n_R	File Size (Bytes)
5	427,996
6	399,666
7	375,412
8	382,314
9	416,166

One major problem with a sub-band audio compressor is the way the high spectral content is dispersed across all transform coefficients. Therefore, whenever a given coefficient (even in the low-frequency CAT space) is heavily decimated (i.e., large T_e) the perceptive effect on the decoded audio is often undesirable. There are smart techniques for minimizing these adverse effects. In one approach, the decimation error in the transform coefficient is added before the final embedded stream encoding is carried out. Such error reduction schemes tend to reduce the compression rates. In the following we examine an alternative approach to using CAT to encode audio data.

Synthetic Digital Audio Generation

The main approach we have, hitherto, followed in deriving CAT basis functions is to impose certain mathematical properties (e.g., orthogonality, smoothness etc.) A more pragmatic approach to digital audio coding will make use of building blocks that are selected on the basis of their audio characteristics, rather than their mathematical attributes. In this section, we show how these synthetic audio building blocks are generated with the same types of rules used to construct the "conventional" CAT filters.

It is desired to generate synthetic digital audio data of duration D seconds consisting of S samples per second with each sample having a maximal value of 2^b. The parameter b represents the number of bits required to encode the specific audio data. For example, if the generated audio is to fit the characteristics of CD-quality stereo music, $S=44100$ and $b=16$. In that case the generated music constitutes one channel of the stereo audio. The other channel can be generated from a different dynamical rule set. For audio music in the mono mode $b=8$. The total number of samples required for a duration of D seconds is $L=SD$.

We desire to generate a data sequence, g_i $(i=0,1,2,...L-1)$, using a cellular automaton lattice of length N. The maximal value of the sequence g is 2^b.

The steps for generating g are the following:

1. Select a dynamical system rule set. The rule set includes

    - The size m of the neighborhood. In the example below $m=3$.

    - The maximum state of the dynamical system must be equal to the maximal value of the sample of the target audio data.

Therefore, $K=2^b$. If the generated data is to include both positive and negative signals, then we set $K=2^{b+1}$. After the CA evolution positive/negative signals are obtained by subtracting 2^b from all generated values.

- The rules W_j $(j=0,1,2,\dots 2^m)$ for evolving the automaton.
- The boundary conditions to be imposed. Most commonly a cyclic condition will be imposed on both boundaries.
- The length, N, of the cellular automaton lattice space.
- The number of steps, T, for evolving the dynamical system is D/N.
- The initial configuration, p_i $(i=0,1,2,\dots N\text{-}1)$, for the cellular automaton. This is a set (total N) of numbers that start the evolution of the CA. The maximal value of this set of numbers is also the maximum state of the automaton.

2. Using the sequence p as the initial configuration, evolve the dynamical system using the rule set selected in (1).

3. Stop the evolution at time $t=T$.

4. To obtain the synthetic audio data, arrange the entire evolved field of the cellular automaton from time $t=1$ to time $t=T$. There are several methods for achieving this arrangement. If a_{jt} is the state of the automaton at node j and time t, two possible arrangements are:

   a. $g_i=a_{jt}$, where $j=i \bmod N$ and $t=(i-j)/N$.

   b. $g_i=a_{jt}$, where $j=(i-t)/N$ and $t= i \bmod T$.

There are infinitely many other algebraic permutations suitable for mapping the field a into the synthetic data g. Figure 4.7 shows a conceptual apparatus, dubbed the *Multi-state Dynamic*

System Audio Processor suitable for generating and playing/recording a synthetic audio given the pertinent rule sets.

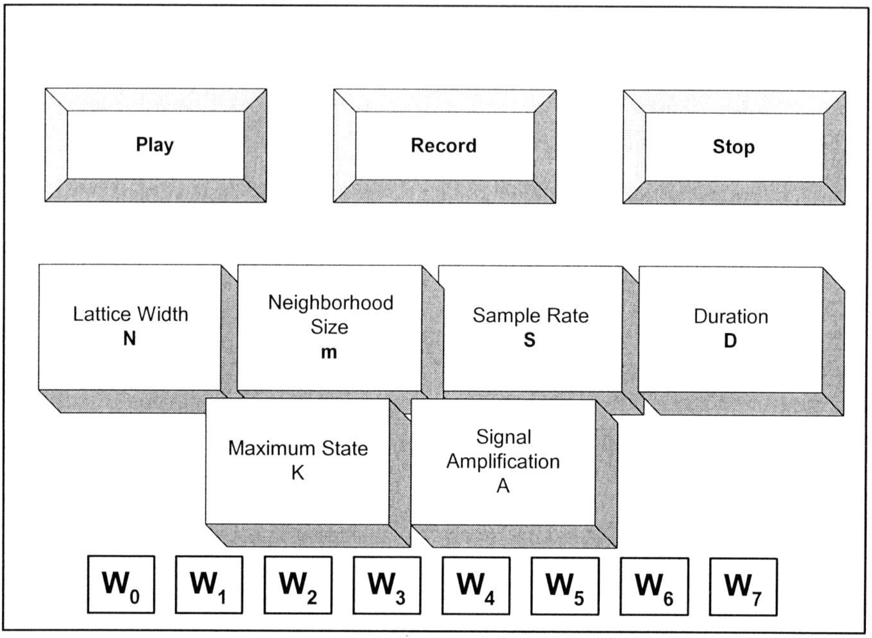

Figure 4.7 *Synthetic multi-state dynamic system audio processor*

Not all CA rules result in acceptable synthetic audio. Obviously the definition of "acceptable" here is closely related to the human perception of the synthetic audio signal. If we consider that pure noise by definition is audio with the spectral energy distributed uniformly over all frequencies, we observe the most "acceptable" synthetic audio will tend to have their energy concentrated in a few frequencies. This energy-concentration property is important in selecting the CA rules for generating synthetic data that will be suitable as good audio building blocks. The steps involved in generating synthetic audio of maximal spectral energy are shown in Figure 4.8. We now outline how the synthetic data can be generated to have pre-specified spectral characteristics.

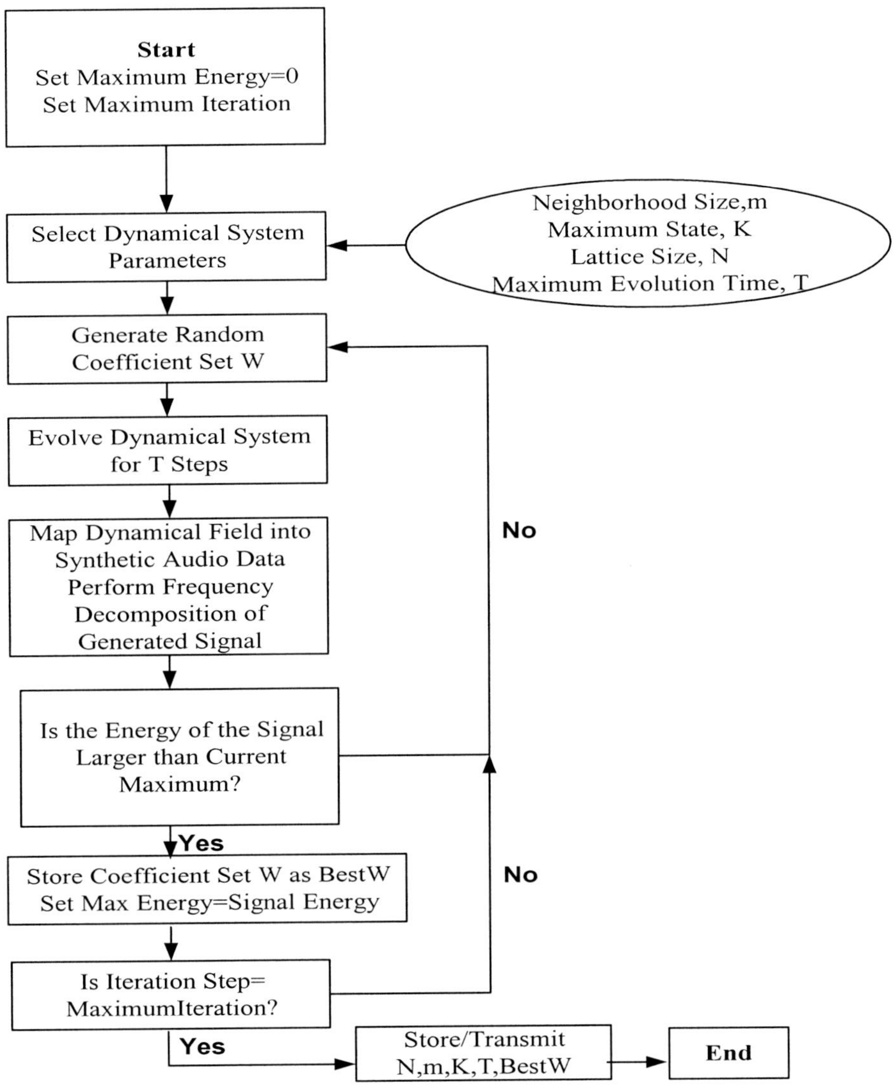

Figure 4.8 *Steps involved in generating digital audio of distinct tonal characteristics*

Synthetic Audio of Specified Frequency

The generated sequence g_i $(i=0,1,2,\ldots\ L-1)$ can be analyzed to determine the audio characteristics. A critical property of an audio sequence is the dominant frequencies. The frequency distribution can be obtained by performing the discrete Fourier transform on the data as:

$$G(n) = \sum_{i=0}^{L-1} g_i e^{2\pi ijn/L} \tag{4.7}$$

where $n=0,1,\ldots L-1$; and $j=\sqrt{(-1)}$. The audio frequency ϕ_n (which is measured in Hertz) is related to the number n and the sampling rate S in the form:

$$\phi_n = \frac{n}{LS} \tag{4.8}$$

To generate audio data of specific frequency distribution (Figure 4.9):

1. Perform the CA generation steps enumerated above.
2. Obtain the discrete Fourier transform of the generated data.
3. Compare the frequency distribution of the generated data with the target distribution. Make a note of the discrepancy between the generated distribution and the target.
4. Select a different set of: 1) coefficients for W-set rule; 2) neighborhood size m; 3) lattice size N.
5. Repeat steps 1-3. Select the rule set that provides the smallest discrepancy.

Figure 4.10 shows the power spectrum of a pair of synthetic audio data with the energy concentrated on the following frequencies: 2,720Hz, 8,225Hz, 13,730Hz, and 19,234Hz. The keys used in the evolution are:

- *N=8,16.*
- *L=65536.*
- W-Rule: See Table 4.4.
- Boundary Condition: Cyclic.

Table 4.4 *Audio encoding W-Rule*

W_0	W_1	W_2	W_3	W_4	W_5	W_6	W_7
113	29	53	11	27	126	26	81

Observe how the change in the base width, *N*, causes a shift in the power spectrum distribution.

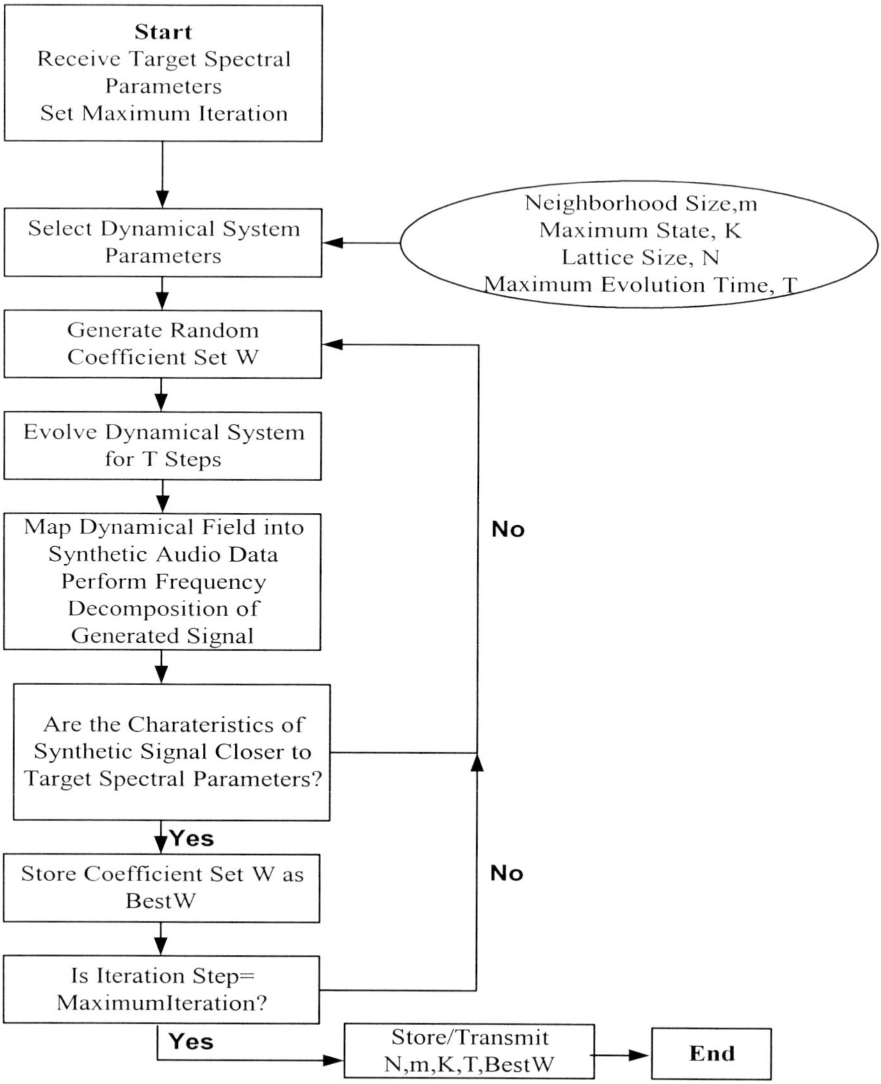

Figure 4.9 *Steps involved in generating digital audio of pre-specified frequency characteristics.*

Digital Audio Coding

Consider the case where a specific audio data sequence, f_i $(i=0,1,2,\ldots L-1)$, is to be encoded. The goal is to find M synthetic CA audio data, g, such that:

$$f_i = \sum_{k=0}^{M-1} c_k g_{ik} \tag{4.9}$$

where g_{ik} is the data generated at point i by k-th synthetic data and c_k is the intensity required in order to correctly encode the given audio sequence.

The encoding parameters are:

1. The W-rule used for the evolution of each of the M synthetic data. For example, if a 3-site neighborhood CA is used for all evolutions, then there are 8 coefficients in each linear rule set.

2. The width N of each automaton.

3. The coefficients c_k that measure the intensity. There are M of these.

To calculate the intensity coefficients, c_k, we write equation (4.9) in the matrix form:

$$\{f\} = [g]\{c\} \tag{4.10}$$

where $\{f\}$ is a column matrix of size L; $\{c\}$ is a column matrix of size M; and $[g]$ is a rectangular matrix of size LM. One approach is to use the least-squares method to determine $\{c\}$ as:

$$\{c\} = [H]^{-1}\{r\}$$
$$H_{mk} = \sum_{i=0}^{L-1} g_{im} g_{ik} \tag{4.11}$$
$$r_m = \sum_{i=0}^{L-1} f_i g_{im}$$

in which $m,k=0, 1, 2, \ldots M-1$. If the group of synthetic CA audio data g_{ik} form an orthogonal set, then it is easy to calculate c_k as:

$$c_k = \frac{1}{\lambda_k} \sum_{i=0}^{L-1} f_i g_{ik} \tag{4.12}$$

where:

$$\lambda_k = \sum_{i=0}^{L-1} g_{ik}^2 \tag{4.13}$$

Notice how the audio building blocks, g_{ik}, are playing the exact role of the CAT basis functions, A_{ik}, used earlier in equation (4.1). In this case the building blocks are synthetic audio data generated using a processor such as the one depicted on Figure 4.7.

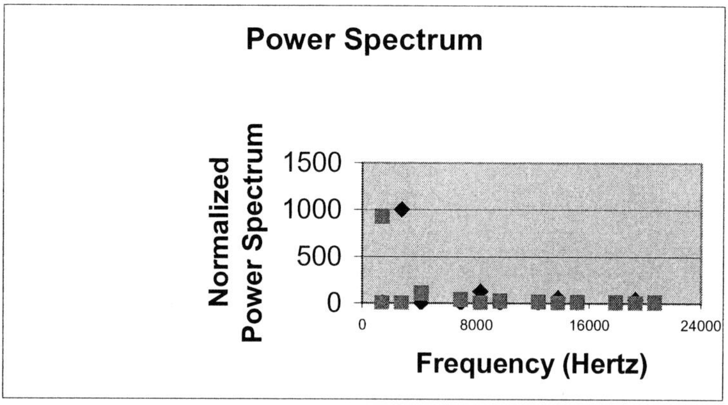

Figure 4.10 *Normalized power,* $(1000P)/P_{\max}$ *, spectrum plots for N=8 (diamonds) and N=16 (squares)*

4.5 Video Compression

The Challenge of Video Coding

At the most primitive level digital video is three-dimensional data (Figure 4.11) consisting of the "flow" of two-dimensional images (the frames) over time. Thirty frames per second (fps) is the standard rate considered to define a fairly good quality video. Eighteen fps will be acceptable for certain situations. High definition video will demand rates of the order of 60 fps.

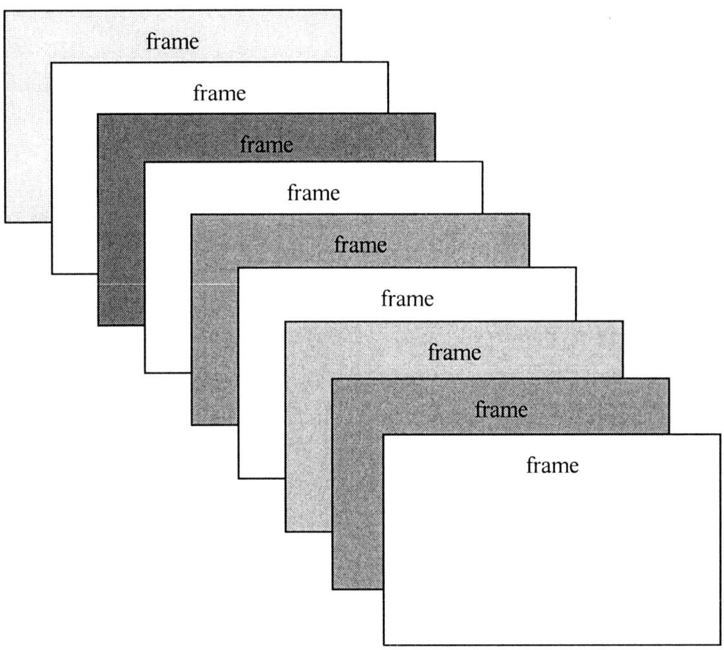

Figure 4.11 *Video as a three-dimensional data block of two-dimensional frames flowing through time*

In a bandwidth-limited environment, the challenge involved in transmitting good quality video is daunting. Consider a video frame consisting of 320x240 pixels. For 24-bit color, each frame will has 3x320x240 = 1,843,200 bits of information. Assuming 30 fps, each second of the video contains 1,843,200x30 = 55,296,000 bits (or 6,912,000 bytes) of data. If this video were to be transmitted through a 56 kilobits per second (kbps) modem, the compression required in order to receive the video in real time is 55,296,000/(56x1,024) = 964:1. Alternatively, to store one hour of this video uncompressed will require a storage space of 6,912,000x60x60 bytes=23Gb. A digital video stream with 640x480 frames will require four times the compression or storage requirement outlined above. Therefore, the need for fast and effective compression is apparent. In the following section we describe three strategies for compressing video data using Cellular Automata Transforms.

The "Video Cube" Approach

This approach uses three-dimensional CAT filters. Video data is treated as three-dimensional information. Each pixel data at the point (i,j) and at a given time t can be represented by:

$$f(i,j,t) = \sum_{k=0}^{N_k-1} \sum_{l=0}^{N_l-1} \sum_{m=0}^{N_m-1} c_{klm} A_{ijtklm} \tag{4.14}$$

where c_{klm} are the transform coefficients, A_{ijtklm} are the CA basis functions and $N_{(.)}$ is the filter size in the respective directions. The filters A_{ijtklm} can be generated in a variety of ways:

- Two-dimensional cellular automata evolved over a specified time. In this approach the basis functions are:

$$A_{ijktklm} = F(a_{ijt} a_{klm}) \tag{4.15}$$

where a represents the states of the CA as they are evolved over time, while F is a mapping function applied as in Classes I and II Schemes described earlier in Chapter 2.

- Products of orthogonal two-dimensional (e.g., 2D, as used for images) and one-dimensional (1D) CAT filters. In this case:

$$A_{ijtklm} = A''_{ijkl} A'_{tm} \tag{4.16}$$

where A'' are 2D bases while A' are 1D basis functions.

- Products of orthogonal 1D CAT filters:

$$A_{ijtklm} = A'_{ik} A'_{jl} A'_{tm} \tag{4.17}$$

The steps involved in the Video Cube encoding approach are:

1. Choose the filter size $N_{(.)}$ in the respective directions. In general we will choose $N_k=N_l=N_m=N$.

2. Select the data for N_m frames.

3. Break each frame into N_k x N_l rectangles.

4. Take each data block (N_k N_l N_m) in sequence. Implement 3D CAT transform on each block. Perform embedded stream encoding. In the 3D case (Figure 4.12), there are eight bands (compare with four bands in 2D) at each resolution level. The coding scheme is also hierarchical. Bands at the coarsest levels will contain the largest coefficients. The coding scheme makes use of the 3-symbol alphabet system: 0 (***YES***), 1 (***NO*** or ***POSV***); and 2 (***NEGV***).

If a target compression ratio C_R is desired, the steps involved in the scheme are the following:

a. Calculate TargetSize = $C_R N_k N_l N_m$.

b. Determine T_{max} = magnitude of coefficient with the largest value throughout all the bands.

c. Set *Threshold* = $2^n > T_{max}$, where n is an integer.

d. Output n. The decoder requires this number.

e. Set OutputSize = 0.

f. Perform the following steps I, II, and III while *OutputSize<TargetSize:*

   I. For each of the sets of data belonging to the 8 Groups, march from the coarsest sub-band to the finest. Determine T_b = maximum coefficient in each sub-band.

   II. If T_b<*Threshold* encode ***YES*** and move onto the next sub-band. Otherwise, encode ***NO*** and proceed to check each coefficient in the sub-band:

   - If the coefficient value is less than *Threshold,* encode ***YES***.

   - Otherwise, encode ***POSV*** if coefficient is positive or ***NEGV*** if it is negative.

   - Decrease the magnitude of the coefficient by *Threshold.*

   III. Set *Threshold* to *Threshold/2*. Return to step (I) if *OutputSize<TargetSize.*

If a target error E_{max} is the goal, the steps involved in the scheme are the following:

a. Determine T_{max}=magnitude of coefficient with the largest value throughout all the bands.

b. Set *Threshold* = $2^n > T_{max}$, where n is an integer.

c. Output n. The decoder requires this number.

d. Perform the following I, II, and III while *Threshold* > E_{max}.

   I. For each of the sets of data belonging to the 8 Groups march from the coarsest sub-band to the finest. Determine T_b = maximum coefficient in each sub-band.

   II. If T_b < *Threshold,* encode ***YES*** and move onto the next sub-band. Otherwise encode ***NO*** and proceed to check each coefficient in the sub-band:

      - If the coefficient value is less than *Threshold* encode ***YES***.

      - Otherwise encode ***POSV*** if coefficient is positive or ***NEGV*** if it is negative.

      - Decrease the magnitude of the coefficient by *Threshold.*

   III. Set *Threshold* to *Threshold/2*. Return to step (I) if *Threshold* > E_{max}.

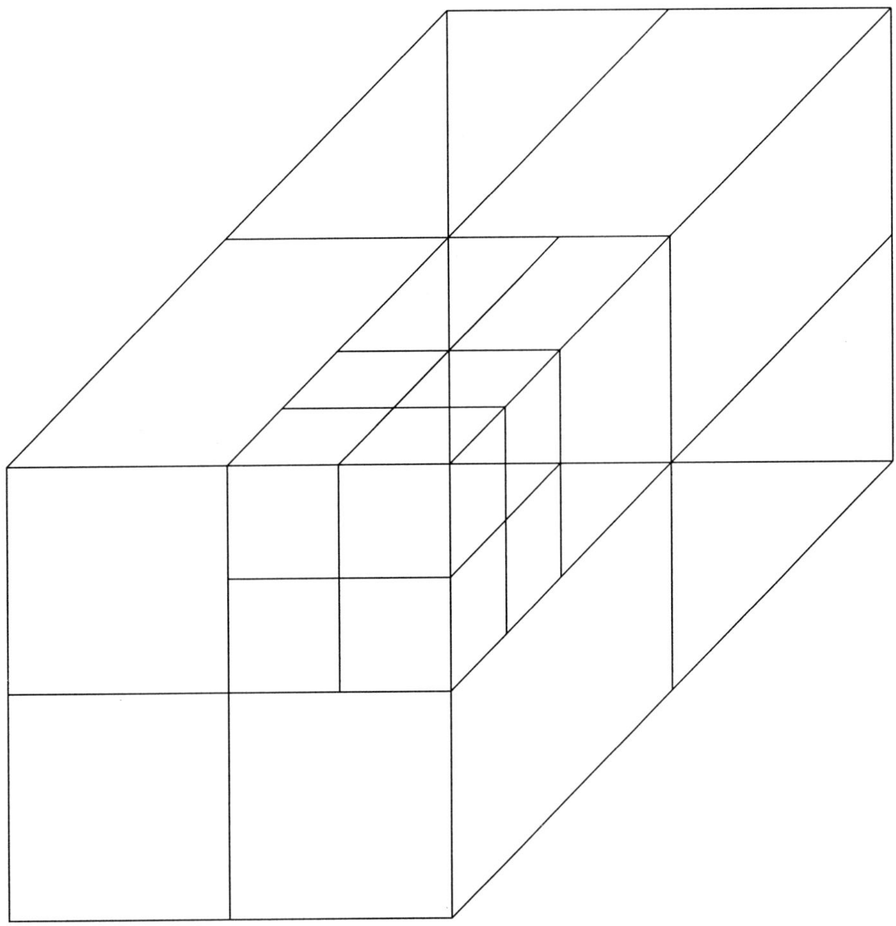

Figure 4.12 *Sub-bands in 3D hierarchical CA transform; each resolution level contains eight bands.*

The same symbol packing strategy described earlier (for image and audio data) is used to store/transmit the decision symbols. The 3D viewpoint has the advantage that the transform process automatically captures all redundancies in the data in time (interframe) and within (intraframe) the frame. The tasks of transforming the video cubes can also be carried out in parallel on a multi-processor machine. Such an approach permits real-time video transmission over a restricted bandwidth channel.

The Sequential Group of Frames Approach

In this approach, a group of video frames is transformed sequentially. The interframe redundancies are captured, via the embedded stream coding, in the frequency space of the transform coefficients. Consider, for example, M frames of video data. Let each frame be transformed in the manner outlined earlier for still images. The transform coefficients are arranged into the various sub-bands. Coefficients in all M frames, belonging to the same sub-band, are grouped together. Let the target error be E_{max}. The steps involved in encoding the coefficients for all M frames are the following:

1. Determine T_{max}=magnitude of coefficient with the largest value in all frames and throughout all the bands.

2. Set *Threshold*= $2^n > T_{max}$, where n is an integer.

3. Output n. The decoder requires this number.

4. Perform Steps i, ii, and iii while *Threshold*>E_{max}:

   i. For each of the sets of data belonging to Groups I, II, III, and IV, in all M frames, march from the coarsest sub-band, to the finest. Determine T_b = maximum coefficient in each sub-band for all M frames.

   ii. If T_b < *Threshold*, encode ***YES*** and move onto the next sub-band. Otherwise, encode **NO** and proceed to check each coefficient in the sub-band group for all M frames:

      a) If the coefficient value is less than *Threshold* encode ***YES***;

      b) Otherwise, encode ***POSV*** if coefficient is positive or ***NEGV*** if it is negative.

      c) Decrease the magnitude of the coefficient by *Threshold*.

iii. Set *Threshold* to *Threshold/2*. Return to step (i) if $Threshold > E_{max}$.

The number M of frames to group together for coding will depend on a number of factors including a) the power of the native processor; b) the degree to which scenes are changing in the video stream; and c) the target error or compression rate.

Reference Frame and Multi-State Predictive Function Approach

The reference frames are encoded using two-dimensional CAT filters. The steps involved are as outlined under the image Compression Scheme Section. The intermediate frames are modeled using a block-based motion estimation scheme that uses one-dimensional predictive functions derived from another set of multi-state dynamical systems. The rules of evolution of the dynamical system and the initial configuration are the key control parameters that determine the characteristics of the generated interpolation functions.

If $R(x,y)$=pixel data at a point (x,y) of the reference frame, the evolution of the pixel data at the point (x,y) in subsequent T "intermediate" frames is obtained as:

$$I(x,y,t) = R(x,y)g_t / \lambda \tag{4.18}$$

where λ is a scaling parameter and the sequence g_t $(t=0,1,2,\ldots\ T\text{-}1)$ is generated using a cellular automaton lattice of length N.

The steps (these are similar to those used for synthetic audio) for generating g are the following:

1. Select a dynamical system rule set. The rule set includes

    - Size m of the neighborhood.

- Maximum state of the dynamical system, $K=2^b$. The maximal value of the sequence g is 2^b. In general, the scaling parameter is $\lambda=2^b$.

- Rules W_j $(j=0,1,2,\ldots 2^m)$ for evolving the automaton.

- Boundary conditions to be imposed. Most commonly cyclic condition will be imposed on both boundaries.

- The length N of the cellular automaton lattice space.

- The number of times T for evolving the dynamical system is the number of intermediate frames before the next reference frame.

- Use the initial configuration, p_i $(i=0,1,2,\ldots N\text{-}1)$, for the cellular automaton. This is a set (total N) of numbers that start the evolution of the CA. The maximal value of this set of numbers is the maximum state of the automaton.

2. Using the sequence p as the initial configuration, evolve the dynamical system using the rule set selected in (1).

3. Stop the evolution at time $t=T$.

4. To obtain the predictive function g we arrange the entire evolved field of the cellular automaton from time $t=1$ to time $t=T$. There are several methods for achieving this arrangement. Using a scheme similar to the one used in synthetic audio generation, with a_{jt}=the state of the automaton at node j and time t, two possible arrangements are:

   a. $g_i=a_{jt}$, where $j=i \bmod N$ and $t=(i-j)/N$.

   b. $g_i=a_{jt}$, where $j=(i-t)/N$ and $t= i \bmod T$.

There are many other permutations for mapping the field a into the function g. The number of predictive functions to use is dictated by the degree of accuracy required to capture the dynamics of the intermediate frames. The most severe condition is when a predictive function is selected for each pixel of the video frame. Obviously such an approach will provide better video quality at the expense of compression size and/or transmission time. The more practical approach is to select the same predictive function for a group of pixels. The predictive functions are encoded by the parameters of the dynamical rules used to generate them.

4.6 Concluding Remarks

Cellular Automata Transforms provide the necessary tools for designing an efficient multimedia compression system. The large library of information building blocks offers flexibility in developing optimal compression algorithms. Furthermore, with CAT, the analyst is offered a choice between using symmetric and asymmetric coding/decoding. The symmetric strategy is the fastest and it is the most appropriate for real-time data compression tasks. For data processing tasks that permit *off-line* encoding (*e.g.*, archiving and CD-ROM production), the asymmetric process is best. In that case, the adaptive power of CA can be exploited while searching for the best transform bases for each data block.

Chapter 5

Data Encryption

5.1 Introduction

For certain kinds of data archival or transmission purposes, the encoding error must be zero. Examples include compression of text files and data encryption applications. There are two approaches for using CAT to encrypt data. The first approach is a straightforward use of the transform process: the plaintext is the input signal, while the transform coefficients constitute the ciphertext. The CA filters have to be integers so the calculations are error free. In the second approach, the plaintext is used as the initial configuration of the cellular automata. The CA rule set is such that the original message is recovered after a fixed length of time, T_f. The ciphertext is the state of the CA at time T_0 such that $0 < T_0 < T_f$.

5.2 Approach I

Consider a one-dimensional sequence of integers f_i $i = 0,1,2,\ldots N-1$. We look for a CA transform consisting of integer transform coefficients c_k, $k = 0,1,2,\cdots N-1$ such that:

$$f_i = \sum_{k=0}^{N-1} c_k A_{ik} \tag{5.1}$$

$$c_k = \sum_{i=0}^{N-1} f_i B_{ik} \tag{5.2}$$

in which the bases A and the inverse bases B must have integer coefficients.

Data is encrypted for security during transmission or in storage systems. The major issue, therefore, is security not compression. With an N-cell, dual-state m-site neighborhood, one-dimensional CA, a code breaker must contend with searching through:

- 2^{2m} rules
- 2^N initial configurations
- 2^{2N} boundary configurations
- Different types of CA bases

The odds against code breakage increase tremendously as the number of states, cellular space, neighborhood and dimensionality increase. If the forward transform bases A are orthogonal, then:

$$B_{ik} = \frac{A_{ik}}{\sum_{j=0}^{N-1} A_{jk}^2}$$

and the coefficients A_{ik} and B_{ik} must be integers for a *floating-point-free* lossless encoding. If the progressive approach is taken, then:

$$c_k = \sum_{i=0}^{N-1} \varepsilon_{ik-1} B_{ik} \quad k = 1,2,3,\cdots N-1 \tag{9.2}$$

$$c_0 = \sum_{i=0}^{N-1} f_i B_{i0}$$

$$\varepsilon_{i0} = f_i - c_0 A_{i0}$$

Windowed progressive bases are the most versatile for lossless encoding.

Implementation

The Approach I CAT-based encryption algorithm has the following features:

- **Symmetric:** The encryption and decryption keys are identical. It is also a **secret-key** algorithm because the sender of the message and the receiver must have sent the key over a secure (preferably different) channel prior to the commencement of the encryption/decryption processes.

- **Block Based:** The plaintext (original message) and ciphertext (encrypted message) are divided into blocks of size N (the length of the cellular space). The implementation here actually uses square blocks of size N x N, although the transform bases are generated from one-dimensional automata. The block size N is included in the key.

- **Variable Key Length:** The key length is easily changed through an *embedded* key generation session. Certain parts (not all) of the keys (*e.g.*, N, initial configuration) can be chosen arbitrarily by the user. However, because the transform bases must form an orthogonal set, the program has internal mechanisms to select the remainder of the keys. The requirement for orthogonality and lossless encoding/decoding obviously puts a restriction on the number of acceptable cellular automata rules. In the second CAT encryption approach we achieve perfect invertibility via a different route.

5.3 Approach II

Consider a data sequence f_i $(i=0,1,2,\dots N\text{-}1)$ derived from a message (a plaintext). The steps for encrypting f using Approach II are as follow:

1. Select a cellular automata rule set. The rule set includes
   - Size m of the neighborhood. In the example below $m=3$.
   - Maximum state K of the CA must be greater than the magnitude of the largest character in the plaintext. In general we select $K=256$.
   - Rules W_j $(j=0,1,2,\dots 2^m)$ for evolving the automaton;
   - Boundary conditions to be imposed. In general, a cyclic condition is imposed on both boundaries .
   - Maximum time T for which the automaton must be evolved. This time is a function of the rule set W_j $(j=0,1,2,\dots 2^m)$.
2. Using the sequence f as the initial configuration, evolve the cellular automaton using the rule set selected in (1).
3. Stop the evolution at time $t=T_f$ when the original message has been recovered. The perfect rule set for the message will recover the message at or before time T. If the original message is not recovered at or before time T, go back to step (1).

4. Select the time $t = T_0$ at which the states of the N cells achieve maximal entropy or disorderliness as measured against the original message.

5. Store or transmit the states (N symbols) of the cellular automata at time T_0 as the ciphertext c_i $(i=0,1,2,\ldots N-1)$.

6. The encryption/decryption keys are:

   - The CA rule set W_j $(j=0, 1, 2, \ldots 2^m)$;

   - The quantities T_0 and $T_d = T_f - T_0$. In the example below $T_d = 1$.

7. To decrypt the message:

   - Use the ciphertext c_i $(i=0,1,2,\ldots N-1)$ as the initial configuration of the CA.

8. With the CA rule set in the encryption/decryption keys, evolve the cellular automata up to the time T_d to recover the original message $f_i (i = 0,1,2,\ldots N-1)$.

An Illustrative Example

Consider the plaintext: *This is a test of cellular automata encryption.*

This message consists of *47 8*-bit characters. To encrypt it we search for a *256*-state cellular automaton.

We pre-select time T=*64* and assume cyclic conditions at the boundaries. That is, we assume the one-dimensional cells are arranged in a circle, thus making the cell *0* to be a neighbor of cell $N-1$. We choose a neighborhood size of 3. With $W_8 = 1$, we have *8* integers W_j, which define the CA evolution rule. The coefficients are shown in the following table.

Table 5.1 *Encryption W-set rule*

W_0	W_1	W_2	W_3	W_4	W_5	W_6	W_7
126	81	84	36	10	4	75	0

The original message was recovered after $T_f = 64$ evolutions of the CA. The ciphertext is:

@_7iNC!_IZL_A~xgd,Y_B__Ul_CS_-U__!VyhS_M2,w_@N

which was obtained from the states of the cells at time $t = 63$. Hence, $T_d = 1$. To decrypt the message, we use the above ciphertext as the initial configuration and evolve the CA for $T_d = 1$ time step.

Although it is described above as a secret-key system, one special feature of Approach II is the possibility of a public-key implementation. A second set of CA key-set, *U*, is required for decoding. The forward evolution (encryption) is carried out, using key-set *W*, to the termination time T_0 at which stage the state of the automaton is the ciphertext. For decryption the different set, *U*, is used to evolve the system for T_d steps with the ciphertext serving as the initial configuration. The key-set W then plays the role of a public key, while *U* is the private key. Obviously the pair of keys (*W*,*U*) must be unique. The search for the keys is more involved than in the secret-key implementation.

General Observations

Unlike most popular block ciphers, CAT encryption (as presented above) operates on bytes, not bits[2]. This may suggest that redundancies in the plaintext would be more apparent in the

[2] There is nothing that theoretically limits CAT encryption to bytes. The algorithm will work on any word length in which the plaintext is presented. Therefore, a bit-based implementation will use a dual-state rule system and receive the plaintext in binary form.

ciphertext. However, there are several ways to alleviate this concern. Like all block ciphers, the strength of CAT encryption can be significantly increased when used in cipher-block-chaining (CBC) mode.

To use CBC, each subsequent block of plaintext is XORed with the previous block of ciphertext before it is encrypted. Thus, for a plaintext block P_i and corresponding ciphertext block C_i:

$$C_i = E_k(C_{i-1} \oplus P_i)$$
$$P_i = C_{i-1} \oplus E_k(C_i)$$

where E_k is the CAT encryption process using the key set k. CBC helps alleviate the redundancy problems in the plaintext. However, this requires the transmission of additional block, the initialization vector V_0, during secure communications. The large block size of CAT encryption means that V_0 might be a sizeable portion of the complete ciphertext.

A good cryptosystem provides both confusion and diffusion. The evolution of the CA serves to confuse the plaintext into an unrecognizable form. This evolution also provides diffusion. Information stored in a cell is spread to the entire neighborhood at every time step. Using a three-site neighborhood CA, a 256-byte block, and a key with period 256, information in a cell will have spread to every other cell in the CA at encryption time.

One interesting question is the characteristics of the CA rule set. For a given rule set to be acceptable for encryption it must be capable of repeating the initial configuration after a finite (and practicable) time. For example, consider a 256-block CA. Theoretically, it may take $256^{256} = 2^{2048}$ time steps for the initial configuration to repeat. That will be too long to make encryption feasible. We have found that the W-rule system gives us CA with small cycles (of the order of the block size).

The weakest rule set is the one using a three-site neighborhood in the linear mode ($W_8 = 1$). A large class of these groups produce CA that require the block size to be an integral power of two to be cyclic. Fortunately, a large family exists that will be cyclic with any size of block. We have observed that rule sets with period 128 have the highest probability of bringing back the initial configuration at any block width. The period 256 rules tend to work with blocks of integral powers of two less than or equal to 256. A side benefit of this is that a plaintext does not have to be padded to a full block width – we can use a dwindling block size method that keeps halving the block size until what is left of the plaintext fits. The plaintext and ciphertext are exactly the same size.

Below we discuss a few methods of attack on the linear small neighborhood CAT encryption. The large neighborhood nonlinear rule sets are infinitely more attack-proof.

- **Brute Force:**

  The key space of the linear three-site neighborhood CAT encryption is 2^{64}. While only a limited number of these are valid for encryption, it is faster to attempt encryption than trying to determine the validity of the rule.

- **Known Plaintext:**

  A cellular automaton system can be thought of as a discretization of a first order partial differential equation, which is locally reversible (*Wolfram*, 1995). Therefore, inverting even a single time step has many solutions. Inverting 64 or 128 time steps is computationally infeasible. The dispersion of information at each time step makes this attack difficult.

- **Chosen Plaintext and Differential Cryptanalysis:**

  As CAT encryption is very different from traditional block ciphers, it is unknown at this time how a chosen plaintext or differential analysis will work. One approach is to choose a plaintext to reduce the possible key space. For example, take a block consisting of only one character: 256 lowercase "a"s. Such a block would go effectively through the CA evolution with four coefficients in the evolution equation. Thus, the key space is reduced to 2^{32}. We can then find all combinations of four numbers that work for that block, and then repeat for a block of all "b"s. This method should quickly indicate which solutions are unacceptable.

- **Reduced Bounds:**

  CAT encryption only involves one round of operations. Therefore the reduced bounds method of attack is non-existent. However, we speculate that linear CAT encryption is a group. Hence, for a specific set of three valid encryption keys k_1, k_2, k_3, and plaintext P:

$$E_{k_1}(E_{k_2}(P)) = E_{k_3}(P)$$

5.4 Concluding Remarks

A CAT-based cryptographic algorithm offers significant advantages over established techniques. These include the ease with which the length L_k of the CA encryption keys can be increased by changing rule parameters such as neighborhood size and maximum state of the automaton. The number of permutations required for a brute-force attack on the key is of the order 2^{Lk}. CAT encryption allows the key selection to be biased toward those basis functions that permit the **avalanche effect** (*Schneier* [1993]), where a one-bit change of the key is supposed to result in a significant change in the ciphertext using the same plaintext, and a one-bit change of the plaintext should yield a significant change in the ciphertext using the same key.

Chapter 6

Solution of Differential and Integral Equations

6.1 Introduction

Differential and integral equations result from the mathematical modeling of processes. For most complex phenomena, the number of independent variables (*e.g.*, spacial coordinates and time) is more than one, and the full description of the process will require more than one dependent variable (*e.g.*, velocity, temperature, displacement, stress). The governing equations for such processes are expressible in the form of partial differential equations (PDEs). These equations indicate the dependence of the process on the plethora of independent variables. For many physical phenomena, PDEs emerge from a continuum viewpoint. The continuum equations derive from statistical averaging of microscopic phenomena. For example, consider a physical process such as the movement of fluid in a container. We can look at the forces acting on a small element of this fluid. This representative elementary volume, REV, should be small enough to allow us to say we are observing the flow at a point, but large enough to make a statistical average of the microscopic events meaningful. When we write statements pertaining to the conservation of mass, momentum and energy for this element, the result will be partial differential equations. Examples include Navier-Stokes equations, Laplace's equation, Poisson equation, etc.

The common approach to solving these PDEs is the use of numerical techniques (*e.g.*, finite difference, finite element, boundary element, finite volume, method of characteristics, etc.) to integrate the pertinent continuum partial differential equations. Tremendous human and financial resources have been expended in the development of these computational methods. Numerous attempts are being made to use cellular automata as modeling tools for complex processes.

6.2 Traditional Cellular Automata Modeling

CA modeling, the basis of the legendary lattice-gas methods, involves the application of the evolving fields of cellular automata to represent the process being studied. The traditional CA modeling approach stands somewhere between the microscopic viewpoint and the continuum approach. In applying CA to a physical problem, an association must be established between known physical quantities of the problem and those calculated as a result of repeated iteration (using the CA rule) from a set of initial conditions (*Gutowitz* [1990]). The process of relating particular CA rules to specific problems is not trivial. However, the association has already been made for a variety of complex nonlinear problems including hydrodynamics (*Frisch et al.*, [1986]), flows in porous media (*Succi et al.*, [1989]; *Rothman* [1990]; *Gunstensen* [1991]), chemical turbulence (*Hartman & Tamayo* [1990]), percolation and nucleation (*Vichniac* [1984]; *Creutz* [1986]), diffusion-controlled reactions (*Canning & Droz* [1990]; *Dab et al.*, [1991]) and dynamical systems (*Cosenza & Neri* [1987]). The above-cited publications have shown that Cellular Automata represent a serious and highly effective tool for studying the *microscopic* character of transport processes, and for solving the associated *macroscopic* (continuum) equations.

CA-based modeling offers advantages both in the representation of the underlying process (physical, chemical, biological, etc.), and in the numerical solution of the problem. For example, the 1986 work of *Frisch, Hasslacher, & Pomeau* (FHP) provided perhaps the first serious effort at using a CA-based method to solve the full Navier-Stokes equations. The fluid is represented by particles of unit mass and momentum moving in a triangular lattice. On reaching a node in the lattice, the particles go through a collision process in which there is a redistribution along the available directions. The total mass and momentum is preserved at each node. Large-scale averages taken from the lattices are solutions to the full Navier-Stokes equations in an asymptotic limit (*Frisch et al.,* [1986,1987]).

The FHP work has had a significant impact on other CA-based modeling. For example, *Rothman* [1988] and *Travis et al.,* [1988] have examined the potential applications of CA methods to the fluid flow in complex porous media. *Wells et al.,* [1991] have investigated CA models for simulating coupled solute transport and chemical reactions at mineral surfaces and in pore networks.

6.3 CA Transform Approach

The CA Transform approach to solving PDEs is more direct than the traditional CA-based/lattice-gas method. We seek to use CA bases much in the same way traditional transforms, such as Fourier and Laplace transforms, are utilized. One advantage in using CA Transforms is the immense number of transform bases. The nice differential and integral properties of these bases can be exploited in obtaining robust solutions to PDEs.

Our principal objective, therefore, is not to model the microscopic process as done in the lattice-gas models. Rather, we want an efficient solver for the continuum equations. In the final analysis, except when called for by the resolution demands of the problem, we will not need millions of cells to solve a given PDE. The number of computational nodes in a CAT-based solution will normally be of the same order of magnitude as those required in traditional solvers based on finite differences and finite element methods. The way the CA differential operators are defined also provides us with robust transform tools for solving nonlinear equations.

Solution Process

The problem domain must be transformed into the cellular automata lattice space. If the domain is regular (*e.g.*, rectangular) the CA lattice space may be a simple discretized version of the physical problem domain. In that case, the character of the partial differential equation remains unchanged. If the domain is irregular, the mapping will be more complex, and the nature of the PDE and its associated initial/boundary conditions may be different in the CA lattice space. In the following presentation, it is assumed that the necessary transformation has been carried out and the PDE is the appropriate equation to be solved in the CA space.

Consider a process governed by the differential equation

$$D[\phi(x,t)] = f(x,t) \tag{6.1}$$

in which ϕ is the dependent variable (*e.g.*, velocity, temperature, pressure, displacement, voltage, current, etc.), D is a differential operator, f is a known forcing function (*e.g.*, effects of sources/sinks and other distributed effects), x is space and t is time. Table 6.1 shows the form of D and the meaning of ϕ for some common physics/engineering problems.

Table 6.1 *Differential operators for some common processes*

Physical Process	Differential Operator, D	Meaning of ϕ
Potential Flows	∇^2	Potential
Heat Conduction	$\nabla.(h\nabla) - \sigma\partial / \partial t$	Temperature
Plate Deformation	∇^4	Deformation
Vibration	$\nabla^4 - a\partial^2 / \partial t^2$	Displacement
Convective-Diffusion	$\partial / \partial t + V.\nabla - \nabla.(h\nabla)$	Pressure
Wave Scattering	$\nabla^2 - k^2$	Wave Field

The CAT-based solution of a differential equation requires the use of cellular automata differential operators. These operators are derived from the CA transform bases described in Part I. Once the CA differential operator is known, the solution is sought by determining the CA transform coefficients associated with the differential bases.

CA Differential Operators

These operators are CA bases that can be used to differentiate a given function. They are constructed from CA basis functions by mapping from the discrete cellular space to the continuous world of the process. In order to illustrate how these CA differential operators are obtained, consider the discretized form of the function $p(x)$, and its first derivative $p'(x)$ along the line $0 \leq x \leq L$. We select N computational nodes on this line, so that the location of the i-th node is $x = iL / (N-1)$.

We expand the function in terms of orthonormal bases A and its derivative using the differential bases D:

$$p_i = \sum_{k=0}^{N-1} c_k A_{ik} \tag{6.2}$$

$$\frac{dp_i}{dx} = p'_i = \frac{1}{\Lambda_i} \sum_{k=0}^{N-1} c_k D_{ik} \tag{6.3}$$

where $D_{ik}=d/di(A_{ik})$ are differential basis functions and Λ_i are scaling parameters. In order to obtain D_{ik} we must have a way of differentiating the basis functions A_{ik}. Our desire is to map A_{ik} into an analytic function so the derivative can be carried out exactly. We have to map from the discrete cellular space to a continuous space. For example, assuming an N-th order polynomial fit, let:

$$A_{ik} = \sum_{j=0}^{N-1} \beta_{jk} (1+i)^j \tag{6.3}$$

where β_{jk} $(j,k=0,1,...N-1)$ are constants obtained by:

$$\{\beta_k\} = [G]^{-1}\{A_k\}$$

in which $G_{ij}=(1+i)^j$. The above must be solved for $k=0,1,2,...N-1$. With the constants thus determined we can obtain:

$$D_{ik} = \sum_{j=0}^{N-1} j\beta_{jk} (1+i)^{j-1}$$

The transform basis functions, shown in Table 6.2 are derived from a Class II, 64-cell, multi-state cellular automata using the W-set rule system. Tables 6.2-6.7 show the results of using the differential basis functions on some common functions. Since the filter size is $N=8$, the points (0, 7, 8, 15) are end points. Notice that the largest errors occur at the end points because the original basis functions are non-overlapping. Overlapping filters automatically incorporate the pertinent continuity constraints required at the end points. This is the primary reason for the usefulness of overlapping filters in the lossy compression of images since some degree of smoothness (sans pixelization) can be maintained at low bit rates.

Table 6.2 *CAT 8x8 orthogonal non-overlap basis functions*

k → i ↓	0	1	2	3	4	5	6	7
0	0.8800	0.1491	0.3428	-0.2695	0.0411	-0.0593	-0.0176	-0.0877
1	0.4400	-0.5556	-0.4083	0.5161	-0.1189	0.2029	-0.0027	0.0967
2	0.1735	0.7331	-0.6173	0.1622	-0.0069	-0.0794	0.0805	0.1104
3	0.0392	-0.3621	-0.4754	-0.5607	0.2448	-0.4875	0.1624	0.0546
4	-0.0041	0.0107	-0.1735	-0.2695	0.5038	0.7608	0.1731	-0.1862
5	0.0021	0.0124	0.0972	0.4682	0.6375	-0.3619	0.0428	-0.4816
6	0.0165	0.0160	0.1458	0.0936	0.5138	-0.0139	-0.2983	0.7851
7	-0.0021	-0.0036	-0.2187	-0.1405	0.0000	0.0383	-0.9198	-0.2913

Table 6.3 *Original function* $f(i) = sin(i)$ *; exact differential* $f'(i) = cos(i)$.

i	CAT Differential	Exact
0	1.00346016	1.00000000
1	0.54128360	0.54030231
2	-0.41700559	-0.41614684
3	-0.98916573	-0.98999250
4	-0.65482921	-0.65364362
5	0.28587772	0.28366219
6	0.95159636	0.96017029
7	0.81687844	0.75390225
8	-0.23789371	-0.14550003
9	-0.89783136	-0.91113026
10	-0.84358606	-0.83907153
11	0.00681705	0.00442570
12	0.84132898	0.84385396
13	0.91112688	0.90744678
14	0.12767500	0.13673722
15	-0.70040373	-0.75968791

Table 6.4 *Original function* $f(i) = (i/15)^3$ *;exact differential* $f'(i) = 3(i/15)^2$

i	CAT Differential	Exact
0	0.00000043	0.00000000
1	0.00073348	0.00073242
2	0.00293163	0.00292969
3	0.00659503	0.00659180
4	0.01172387	0.01171875
5	0.01831847	0.01831055
6	0.02637927	0.02636719
7	0.03590680	0.03588867
8	0.04688949	0.04687500
9	0.05936193	0.05932617
10	0.07330860	0.07324219
11	0.08872767	0.08862305
12	0.10561279	0.10546875
13	0.12395098	0.12377930
14	0.14371967	0.14355469
15	0.16488308	0.16479492

Table 6.5 *Original function* $f(i) = \log(1+i)$; *exact differential* $f'(i) = 1.0/(1+i)$

i	CAT Differential	Exact
0	0.95974775	1.00000000
1	0.50342467	0.50000000
2	0.33282341	0.33333333
3	0.25075953	0.25000000
4	0.20023316	0.20000000
5	0.16763469	0.16666667
6	0.14215313	0.14285714
7	0.13138732	0.12500000
8	0.11120955	0.11111111
9	0.10025264	0.10000000
10	0.09138133	0.09090909
11	0.08407202	0.08333333
12	0.07790951	0.07692308
13	0.07251541	0.07142857
14	0.06748844	0.06666667
15	0.06235679	0.06250000

Table 6.6 *Original function* $f(i) = \exp(-i)$ *; exact differential* $f'(i) = -\exp(-i)$

i	CAT Differential	Exact
0	0.99260880	1.00000000
1	0.36878606	0.36787944
2	0.13504870	0.13533528
3	0.04989474	0.04978707
4	0.01812182	0.01831564
5	0.00682391	0.00673795
6	0.00176014	0.00247875
7	0.00409579	0.00091188
8	0.00033298	0.00033546
9	0.00012371	0.00012341
10	0.00004530	0.00004540
11	0.00001674	0.00001670
12	0.00000608	0.00000614
13	0.00000229	0.00000226
14	0.00000059	0.00000083
15	0.00000137	0.00000031

Table 6.7 *Original function* $f(i) = i\cos(i)$*; exact differential* $f'(i) = \cos(i) - i\sin(i)$

i	CAT Differential	Exact
0	0.73853237	1.00000000
1	-0.25037267	-0.30116868
2	-2.25669809	-2.23474169
3	-1.39831986	-1.41335252
4	2.35519770	2.37356636
5	5.11053817	5.07828356
6	2.53456580	2.63666328
7	-3.10574526	-3.84500394
8	-8.63204849	-8.06036601
9	-4.55340066	-4.62019663
10	4.58559693	4.60113958
11	11.01022228	11.00431797
12	7.28339932	7.28272897
13	-4.55917990	-4.55472470
14	-13.68438531	-13.73176576
15	10.94743103	-10.51400552

The procedure for solving a differential equation using CAT filters is quite straightforward:

1. Evolve the CA filters as outlined in Chapter 3.

2. Map the filters from the discrete space to a continuous space (for example using polynomial map used above).

3. Write the solution to the problem in the form of a CAT series. This solution will contain the transform coefficients as unknowns.

4. Apply this solution to the governing differential equation. The CA differential operators will emerge at this stage.

5. Use the orthogonality property to eliminate the transform coefficients. The unknowns will be the dependent variables at the cells.

6. Introduce the boundary and/or initial conditions.

7. Solve the ensuing system of equations to determine the unknown quantities.

Numerical Examples

One-Dimensional Example

We illustrate the above solution steps with a simple example. Consider the one-dimensional heat conduction problem in a rod of length L:

$$\begin{aligned} &\frac{d^2\Phi}{dX^2} = F(X) \\ &\Phi(0) = T_0 \\ &\Phi(L) = T_L \end{aligned} \tag{6.4}$$

Where $F(X)$ is a heat recharge term. We render this dimensionless by defining:

$$\begin{aligned} x &= X/L \\ \phi &= (\Phi - T_0)/(T_L - T_0) \\ f &= FL^2/(T_L - T_0) \end{aligned}$$

so as to obtain:

$$\frac{d^2\phi}{dx^2} = f(x)$$
$$\phi(0) = 0 \tag{6.5}$$
$$\phi(1) = 1$$

We divide the region $0 \le x \le 1$ into N segments. Let $i=(N-1)x$. The solution is the form:

$$\phi_i = \sum_{k=0}^{N-1} c_k A_{ik} \tag{6.6}$$

We use the above in the governing equation to obtain:

$$\sum_{k=0}^{N-1} c_k D_{ik} = \frac{f(x)}{(N-1)^2} \tag{6.7}$$

where:

$$D_{ik} = \frac{d^2 A_{ik}}{di^2} \tag{6.8}$$

Since the basis functions A_{ik} are orthonormal, we can write:

$$c_k = \sum_{j=0}^{N-1} \phi_j A_{jk} \tag{6.9}$$

which when used in equation (6.7) results in:

$$\sum_{k=0}^{N-1} \left(\sum_{j=0}^{N-1} \phi_j A_{kj}\right) D_{ik} = \frac{f(x)}{(N-1)^2} \tag{6.10}$$

or:

$$\sum_{j=0}^{N-1} \phi_j H_{ij} = \frac{f(x)}{(N-1)^2} \tag{6.11}$$

in which:

$$H_{ij} = \sum_{k=0}^{N-1} A_{jk} D_{ik} \tag{6.12}$$

Using the orthonormal basis functions shown in Table 6.2, the associated second order differential basis functions D_{ik} are shown in Table 6.8. The solution matrix H_{ij} is shown in Table 6.9.

Table 6.8 *Second order differential basis functions, D_{ik}*

k → i ↓	0	1	2	3	4	5	6	7
0	33.3752	-39.5115	-49.4402	37.6686	12.7953	13.3172	20.3110	-7.9155
1	-1.9037	4.8928	2.1228	-3.0780	-1.4221	-1.1173	-1.6839	0.4345
2	-0.7540	0.4238	2.3005	0.4367	0.2100	0.5677	-0.0088	0.2637
3	1.3063	-2.1390	-1.4742	1.3112	1.1154	-0.1189	0.7316	0.0385
4	-0.6519	1.4569	0.2313	-2.1934	-0.6357	-1.6015	-1.2464	0.0140
5	-0.0856	-0.2127	0.5729	1.2390	-1.0897	1.4952	0.0012	-1.1580
6	1.9026	-3.3988	-1.8463	3.8727	0.8275	2.0793	3.0522	0.1803
7	-22.5525	41.6712	18.7477	-53.1871	-5.4522	-35.6749	-15.9744	19.3007

Table 6.9 *Solution matrix, H_{ij}*

k → i ↓	0	1	2	3	4	5	6	7
0	5.2115	-22.3016	43.9536	-52.7271	41.0042	-20.1022	5.6618	-0.7001
1	0.7000	-0.3889	-2.6999	4.7499	-3.7221	1.7999	-0.5000	0.0611
2	-0.0611	1.1887	-2.0995	0.7216	0.4729	-0.3003	0.0890	-0.0111
3	0.0112	-0.1504	1.5010	-2.7235	1.5012	-0.1506	0.0113	-0.0000
4	0.0001	0.0105	-0.1485	1.4978	-2.7202	1.4989	-0.1497	0.0111
5	-0.0109	0.0878	-0.2972	0.4679	0.7260	-2.1021	1.1896	-0.0612
6	0.0614	-0.5011	1.8028	-3.7269	4.7538	-2.7021	-0.3881	0.6999
7	-0.7003	5.6636	-20.1065	41.0084	-52.7306	43.9547	-22.3012	5.2113

The solution process need not involve an explicit inversion of the solution matrix. In fact, we favor an iterative solution of the form:

$$\phi_i^{n+1} = \theta\phi_i^n + (1-\theta)\left(f_i - \sum_{j=0;\, j\neq i}^{N-1} H_{ij}\phi_j^n \right) / H_{ii} \qquad (6.13)$$

Where n is the iteration step, and $0 < \theta < 1$ is a relaxation parameter. The above scheme also allows an easy incorporation of the boundary condition. Table 6.10 shows the solution for the case $f(x) = x^2$. The accuracy of the CAT solution is clearly evident.

Table 6.10 *Solution for* $f(x) = x^2$ *; exact* $\phi(x) = x(x^3 + 11)/12$

i	X	CAT Solution	Exact
0	0.00000000	0.00000000	0.00000000
1	0.14285714	0.13108258	0.13098709
2	0.28571429	0.26261969	0.26246009
3	0.42857143	0.39585042	0.39566847
4	0.57142857	0.53285219	0.53269471
5	0.71428571	0.67654744	0.67645426
6	0.85714286	0.83071417	0.83069554
7	1.00000000	1.00000000	1.00000000

An easy extension of the solution process is the case of a heat leakage problem:

$$\begin{aligned} &\frac{d^2\phi}{dx^2} - \lambda^2\phi = f(x) \\ &\phi(0) = 0 \\ &\phi(1) = 1 \end{aligned} \tag{6.14}$$

Where λ is the leakage parameter. The iterative solution involves simply augmenting the forcing function *f(x)* with the leakage term:

$$\phi_i^{n+1} = \theta\phi_i^n + (1-\theta)\left(f_i + \lambda^2\phi_i^n - \sum_{j=0;\,j\neq i}^{N-1} H_{ij}\phi_j^n \right) / H_{ii} \tag{6.15}$$

The CAT solution is shown on Table 6.11.

Table 6.11 *Solution for* $\lambda = 1$ *and* $f(x) = x^2$;
exact $\phi(x) = (11\sinh(\lambda x)/\sinh(\lambda) + x^4)/12$

i	x	CAT Solution	Exact
0	0.00000000	0.00000000	0.00000000
1	0.14285714	0.11157523	0.11184386
2	0.28571429	0.22589054	0.22645934
3	0.42857143	0.34652720	0.34742833
4	0.57142857	0.47803389	0.47926015
5	0.71428571	0.62602391	0.62744109
6	0.85714286	0.79729875	0.79848704
7	1.00000000	1.00000000	1.00000000

Problems involving other derivatives can be determined the same way the first and second order derivatives have been obtained above. For example, the q-th derivative is:

$$f_i^q = \sum_{k=0}^{N-1} c_k D_{ik}^q \tag{6.16}$$

where D_{ik}^q are the q-th order differential operator.

Two-Dimensional Example

The earlier numerical example is based on a one-dimensional differential equation problem. The following example shows how a one-dimensional CAT basis function can be used in solving a multi-dimensional problem. Consider the two-dimensional Poisson equation:

$$\frac{\partial^2\phi}{\partial x^2}+\frac{\partial^2\phi}{\partial y^2}=f(x,y) \tag{6.17}$$

With the Dirichlet condition $\phi=\phi_b$ on the boundary $\Gamma=\Gamma_b$. We divide the region into an $N\times N$ grid and assume the solution in the form:

$$\phi_{ij}=\sum_{m=0}^{N-1}\sum_{n=0}^{N-1}c_{mn}A_{im}A_{jn} \tag{6.18}$$

where $x=j/(N-1), y=i/(N-1)$; A are orthonormal one-dimensional CA basis functions; and c_{mn} are the transform coefficients. Given the second-order differential bases:

$$D_{im}=\frac{d^2A_{im}}{di^2}$$

we can transform the governing equation into the form

$$\sum_{m=0}^{N-1}\sum_{n=0}^{N-1}c_{mn}\left(D_{im}A_{jn}+A_{im}D_{jn}\right)=f_{ij}/(N-1)^2 \tag{6.19}$$

Since the basis functions are orthonormal, the transform coefficients can be obtained in the form:

$$c_{mn}=\sum_{k=0}^{N-1}\sum_{l=0}^{N-1}\phi_{kl}A_{mk}A_{nl} \tag{6.20}$$

which when used in equation (6.19) results in[3]:

$$\sum_{k=0}^{N-1}\sum_{l=0}^{N-1}\phi_{kl}H_{ijkl} = f_{ij}/(N-1)^2 \tag{6.21}$$

in which the solution matrix is

$$H_{ijkl} = \sum_{m=0}^{N-1}\sum_{n=0}^{N-1} A_{mk}A_{nl}\left(D_{im}A_{jn} + A_{im}D_{jn}\right) \tag{6.22}$$

Similar derivations can be performed in higher dimensions. Using the CA basis functions shown in Table 6.12, the result of a numerical solution of the Poisson problem, with $f(x,y) = x/(1+y)^2$ on a unit square, is presented in Table 6.13. The imposed condition on the boundary is

$$\phi\left(x_b, y_b\right) = \cos(x_b)\sinh(y_b) + x_b\log(1+y_b) \tag{6.23}$$

[3] Note that equation (6.21) provides a means for carrying out the Laplacean differentiation of a given two-dimensional function. Such differentiation can be used for digital image edge detection. Hence, if an NxN image data is represented by ϕ, then the Laplacean is $f_{ij} = \left(N-1\right)^2\sum_{k=0}^{N-1}\sum_{l=0}^{N-1}\phi_{kl}H_{ijkl}$.

Table 6.12 *Basis functions A used to construct Poisson solver*

j → i ↓	0	1	2	3	4	5	6	7
0	0.8975	-0.1787	0.2661	0.1212	-0.1107	-0.1533	0.0728	0.1894
1	-0.4177	0.6143	-0.3826	-0.3094	0.1909	0.2578	-0.1306	-0.2931
2	-0.1376	-0.7148	-0.6087	-0.0204	0.1496	0.1773	-0.0765	-0.1982
3	-0.0075	0.2814	-0.5446	0.6891	-0.0921	-0.2599	0.0929	0.2538
4	0.0223	-0.0049	-0.3230	-0.6090	-0.3918	-0.5242	0.2355	0.2009
5	0.0016	0.0153	-0.0762	-0.0093	-0.6068	0.7111	0.2090	0.2761
6	-0.0197	-0.0014	0.0629	0.1987	-0.5945	-0.1702	-0.1289	-0.7463
7	0.0077	-0.0110	-0.0377	-0.0609	-0.2123	-0.0385	-0.9205	0.3173

Table 6.13 *Solution at internal nodes for two-dimensional Poisson problem*

i	j	x	y	CAT Solution	Exact	Error
1	1	0.14285714	0.14285714	0.15200889	0.16095926	0.0090
1	2	0.28571429	0.14285714	0.16343881	0.17568431	0.0122
1	3	0.42857143	0.14285714	0.17655744	0.18760735	0.0110
1	4	0.57142857	0.14285714	0.18722981	0.19687410	0.0096
1	5	0.71428571	0.14285714	0.19292361	0.20368441	0.0108
1	6	0.85714286	0.14285714	0.20465116	0.20828818	0.0036
2	1	0.14285714	0.28571429	0.31749386	0.32256926	0.0051
2	2	0.28571429	0.28571429	0.34308845	0.34968065	0.0066
2	3	0.42857143	0.28571429	0.36313766	0.37113074	0.0080
2	4	0.57142857	0.28571429	0.37715340	0.38721394	0.0101
2	5	0.71428571	0.28571429	0.38668853	0.39833406	0.0116
2	6	0.85714286	0.28571429	0.39996958	0.40499597	0.0050
3	1	0.14285714	0.42857143	0.50220436	0.48826492	0.0139
3	2	0.28571429	0.42857143	0.54361087	0.52580830	0.0178
3	3	0.42857143	0.42857143	0.56381714	0.55471535	0.0091
3	4	0.57142857	0.42857143	0.57390827	0.57543521	0.0015
3	5	0.71428571	0.42857143	0.58185989	0.58858387	0.0067
3	6	0.85714286	0.42857143	0.58658636	0.59493153	0.0083
4	1	0.14285714	0.57142857	0.68680215	0.66146472	0.0253
4	2	0.28571429	0.57142857	0.74138319	0.70773020	0.0337
4	3	0.42857143	0.57142857	0.76497257	0.74220774	0.0228
4	4	0.57142857	0.57142857	0.77502900	0.76551044	0.0095
4	5	0.71428571	0.57142857	0.78053194	0.77847903	0.0021
4	6	0.85714286	0.57142857	0.77636582	0.78216481	0.0058
5	1	0.14285714	0.71428571	0.85831535	0.84568126	0.0126
5	2	0.28571429	0.71428571	0.92232317	0.89910909	0.0232
5	3	0.42857143	0.71428571	0.96293533	0.93735643	0.0256
5	4	0.57142857	0.71428571	0.98041761	0.96121281	0.0192
5	5	0.71428571	0.71428571	0.98052657	0.97176093	0.0088
5	6	0.85714286	0.71428571	0.96995878	0.97035464	0.0004
6	1	0.14285714	0.85714286	1.04613495	1.04461635	0.0015
6	2	0.28571429	0.85714286	1.11490512	1.10372914	0.0112
6	3	0.42857143	0.85714286	1.16745543	1.14395855	0.0235
6	4	0.57142857	0.85714286	1.19029403	1.16628669	0.0240
6	5	0.71428571	0.85714286	1.18278480	1.17206035	0.0107
6	6	0.85714286	0.85714286	1.16285241	1.16296362	0.0001

Nonlinear Formulation

The solution process outlined for the Poisson equation provides the necessary foundation for using CAT for a class of nonlinear problems. Consider the two-dimensional flow of an incompressible fluid. Given the dimensionless velocity $v = (u, v)$ and the dimensionless pressure p, the governing equations are the incompressible Navier-Stokes equations:

$$\begin{aligned}
&\frac{\partial u}{\partial x} + \frac{\partial v}{\partial y} = 0 \\
&u\frac{\partial u}{\partial x} + v\frac{\partial u}{\partial y} = -\frac{1}{\rho}\frac{\partial p}{\partial x} + \upsilon\left(\frac{\partial^2 u}{\partial x^2} + \frac{\partial^2 u}{\partial y^2}\right) \\
&u\frac{\partial v}{\partial x} + v\frac{\partial v}{\partial y} = -\frac{1}{\rho}\frac{\partial p}{\partial y} + \upsilon\left(\frac{\partial^2 v}{\partial x^2} + \frac{\partial^2 v}{\partial y^2}\right)
\end{aligned} \tag{6.24}$$

in which v is the kinematic viscosity. One approach to solving these are the so-called Vorticity-Stream function Method. Let

$$\begin{aligned}
\zeta &= \frac{\partial v}{\partial x} - \frac{\partial u}{\partial y} \\
u &= \frac{\partial \Psi}{\partial y} \\
v &= -\frac{\partial \Psi}{\partial x}
\end{aligned} \tag{6.25}$$

where ζ vorticity and Ψ is the stream function. When the above definitions are used in equation (6.23) we obtain the following system of Poisson equations:

$$\frac{\partial^2 \Psi}{\partial x^2} + \frac{\partial^2 \Psi}{\partial y^2} = -\zeta$$

$$\frac{\partial^2 \zeta}{\partial x^2} + \frac{\partial^2 \zeta}{\partial y^2} = R_e\left(u\frac{\partial \zeta}{\partial x} + v\frac{\partial \zeta}{\partial y}\right) \tag{6.26}$$

$$\frac{\partial^2 p}{\partial x^2} + \frac{\partial^2 p}{\partial y^2} = -2\left[\left(\frac{\partial v}{\partial x}\right)\left(\frac{\partial u}{\partial y}\right) - \left(\frac{\partial u}{\partial x}\right)\left(\frac{\partial v}{\partial y}\right)\right]$$

in which R_e is the Reynolds Number. The above equations are the respective equations for the stream function, vorticity and pressure. It is not necessary to determine pressure solution until the steam function and vorticity distributions are known.

The CAT solution process will be iterative and consist of the following steps:

1. Divide the flow region into a grid of size *N x N*.
2. Start with some initial distribution for Ψ and ζ for all grid points.
3. Use Ψ to compute the velocity field (u,v) via equation (6.25).
4. Obtain a new distribution for ζ by differentiating the velocities.
5. With ζ as the forcing function, solve equation (6.26) to obtain a new distribution for Ψ.
6. Find a new velocity field from the Ψ distribution.
7. Return to Step 3 if there is no convergence.

Treatment of Time

Derivatives in time are handled the same way spatial derivatives are obtained above. Use is made of time-accurate, transient CA differential operators. Hence, for time-dependent function *f(t)*, discretized into $f_n(n = 0,1,2,\ldots N_T - 1)$ we write:

$$\frac{df_n}{dt} = \frac{1}{\Lambda_n} \sum_{l=0}^{N_T-1} c_l D_{nl} \tag{6.27}$$

where Λ_n are time-based scaling factors, while D_{nl} are first-order CA differential operators.

If $f = f(x,t)$,then the discretization will result in

$$f_{in}(i = 0,1,2,\ldots N-1; n = 0,1,2,\ldots N_T - 1)$$

and:

$$\frac{\partial^2 f_{in}}{\partial x \partial t} = \frac{1}{\Lambda_{in}} \sum_{l=0}^{N_T-1} \sum_{i=0}^{N-1} c_{kl} D_{ik} D_{nl} \tag{6.28}$$

In general, for an order $p \geq 2$ derivative in time, and order $q \geq 2$ in space, we have high-order CA differential operators D^p and D^q in the form:

$$\frac{\partial^{p+q} f_{in}}{\partial^p t \partial^q x} = \frac{1}{\Lambda_{in}} \sum_{l=0}^{N_T-1} \sum_{i=0}^{N-1} c_{kl} D_{ik}^q D_{nl}^p \tag{6.29}$$

6.4 Integral Equations

Many solution schemes in mathematical physics are based on the conversion of PDEs to integral equations. For example, using the Greens identities, many elliptic PDEs can be converted into boundary integral equations. Other PDEs may require the use of some reciprocal relations (*e.g.*, Betti's formula in linear elastostatics). In some instances, the physical problem is cast directly in the form of an integral equation by summing the effects of distributed fictitious point actions (*e.g.*, sources and dipoles) on the problem boundary.

Consider the integral equation of the second kind

$$\alpha f(x) = \int_a^b K(x,\zeta) f(\zeta) d\zeta + g(x) \tag{6.30}$$

where α, a, and b are constants, f is to be determined, g is known and K is the kernel. When α=0, the above degenerates into an integral equation of the first kind. Table 6.14 shows the kernel (or the free-space Green's function) for the partial differential equations associated with some common physical processes.

Table 6.14 *Kernel's function for converting some differential equations into integral equations*

Differential Operator	Kernel	Dimensions	Remarks
∇^2	$\ln(r)/2\pi$	2	$r=\lvert x-\varsigma\rvert$
∇^2	$-1/4\pi r$	3	
$\nabla^2-\lambda^2$	$-K_0(\lambda r)/2\pi$	2	K_0=Zeroth order modified Bessel Function
$\nabla^2-\lambda^2$	$-\exp(-\lambda r)/4\pi r$	3	
$\nabla^2+\lambda^2$	$Y_0(\lambda r)/4\pi$	2	Y_0=Zeroth order Bessel Function
$\nabla^2+\lambda^2$	$-\cos(\lambda r)/4\pi r$	3	
∇^4	$r^2\ln r/8\pi$	2	

In most applications, matrix equations resulting from the numerical discretization of equation (6.30) are fully populated (or dense). However, by transforming the salient quantities f, g and/or K into a CA space, sparse coefficient matrices can be obtained regardless of the character of the kernel functions. We present two approaches: decomposition of kernel functions and decomposition of variables.

Decomposition of Kernel Functions

This approach is in line with the so-called Nyström (see *e.g., Alpert et al.* [1993]). Using an N-point quadrature rule, we can write:

$$\int_a^b K(x,\zeta)f(\zeta)d\zeta = \sum_{j=0}^{N-1} w_j K(x,\zeta_j)f(\zeta_j) \tag{6.31}$$

where w are the weights of the quadrature formula. Using equation (6.31) in (6.30) we have for the i-th point:

$$\alpha f_i = \sum_{j=0}^{N-1} w_j K_{ij} f_j + g_i \quad i = 0,1,2,\dots N-1 \tag{6.32}$$

where $f_i = f(x_i)$, $g_i = g(x_i)$, and $K_{ij} = K(x_i, z_j)$. We represent the discrete kernel functions by orthogonal CA bases A in the form:

$$K_{ij} = \sum_{k=0}^{N-1}\sum_{l=0}^{N-1} c_{kl} A_{ijkl} \tag{6.33}$$

$$c_{kl} = \sum_{i=0}^{N-1}\sum_{j=0}^{N-1} K_{ij} B_{ijkl} \tag{6.34}$$

in which B is the inverse of A. When equation (6.33) is used in equation (6.32) the result is:

$$af_i = \sum_{j=0}^{N-1}\sum_{k=0}^{N-1}\sum_{l=0}^{N-1} w_j c_{kl} A_{ijkl} f_j + g_i \quad i = 0,1,2,\dots N-1 \tag{6.35}$$

By using CA bases A that maximize the number of zero or non-significant transform coefficients c (compare with the requirements for image compression discussed in Chapter 4), the result is a sparse matrix for solving for f_i $(i = 0,1,2,\ldots N-1)$. The system of equation (6.35) can be written in the matrix form:

$$[H]\{f\} = \{g\} \tag{6.36}$$

in which:

$$H_{ij} = \delta_{ij}\alpha - \sum_{k=0}^{N-1}\sum_{l=0}^{N-1} c_{kl} A_{ijkl} \qquad i = 0,1,2,\ldots N-1$$

Decomposition of Variables

This is similar to the Galerkin (*Alpert et al.* [1993]) approach. We write:

$$f_i = \sum_{k=0}^{N-1} c_k A_{ik} \tag{6.37}$$

$$c_k = \sum_{i=0}^{N-1} f_i B_{ik}$$

which when used in equation (6.30) results in:

$$\sum_{k=0}^{N-1} c_k(\alpha A_{ik} - K_{ik}) = g_i \quad i = 0,1,2,\ldots N-1 \tag{6.38}$$

where:

$$K_{ik} = \int_a^b K(x_i,\zeta) A_{ik} d\zeta$$

Again, by favoring CA bases A which result in a large number of zero/near-zero transform coefficients, the matrix equation resulting from equation (6.38) can be made as sparse as possible. Equation (6.38) can be written in the matrix form:

$$[H]\{c\} = \{g\} \tag{6.39}$$

in which the coefficients:

$$H_{ij} = \alpha A_{ij} - K_{ij}$$

The kernel-decomposition approach involves the evaluation of fewer terms than the Galerkin technique. The Galerkin approach requires the use of one-dimensional basis functions, while the kernel-decomposition method requires two-dimensional bases. On balance, the kernel-decomposition approach holds a computational advantage over the Galerkin method.

6.5 Concluding Remarks

The fundamental concepts involved in the application of CA transforms to partial differential equations and integral equations have been outlined. The derivations show the versatility of these transforms in obtaining robust solutions to differential and integral equations. The CAT solution approach does not require the millions of computational cells used in the traditional methods. CAT-based solutions incorporate a huge library of rules which can be adapted to specific problem features.

Appendix A

```
int EvolveCellularAutomata(int *a)
{
      int i,j,seed,p,D=0,Nz=NeighborhoodSize-
1,Residual;
      for (i=0;i<RuleSize;i++)
      {
            seed=1;p=1 << Nz;Residual=i;
            for (j=Nz;j>=0;j--)
            {
                  if (Residual >= p)
                  {
                        seed *= a[j];
                        Residual -= p;
                  }
                  if (seed == 0) break;
                  p >>= 1;
            }
            D += (seed*W[i]);
      }
      return (D % STATE);
}
```

Program A.1 *A C-code for evolving one-dimensional CA for a given STATE and NeighborhoodSize. Vector {a} represents the states of the cells in the neighborhood. RuleSize=*$2^{NeighborhoodSize}$.

Appendix B

Program B.1 is the fast transform for a one-dimensional dual-coefficient CAT basis function. The input data f is a vector of eight values. The routine returns the transform in f. The transform is fully symmetric.

```
void CATransform(double *f)
{
        double P[15],lambda=sqrt(8);
        P[0]=f[0]+f[1];
        P[1]=f[0]-f[1];
        P[2]=f[2]+f[3];
        P[3]=f[2]-f[3];
        P[4]=f[4]+f[5];
        P[5]=f[4]-f[5];
        P[6]=f[6]+f[7];
        P[7]=f[6]-f[7];
        P[8]=P[0]-P[4];
        P[9]=-P[2]-P[6];
        P[10]=P[1]-P[5];
        P[11]=-P[3]-P[7];
        P[12]=-P[0]-P[4];
        P[13]=P[2]-P[6];
        P[14]=-P[1]-P[5];
        P[15]=P[3]-P[7];
        f[0]=P[8]+P[9];
        f[1]=P[10]+P[11];
        f[2]=P[12]+P[13];
        f[3]=P[14]+P[15];
        f[4]=-P[8]+P[9];
        f[5]=-P[10]+P[11];
        f[6]=P[12]-P[13];
        f[7]=P[14]-P[15];
        for (i=0;i<8;i++) f[i] /= lambda;

}
```

Program B.1 *Fast one-dimensional, dual-coefficient CA Transform*

Program B.2 is the fast transform for a two-dimensional, dual-coefficient CAT basis function (See Figure 3.2). The input data f is a vector of 64 values. The routine returns the transform coefficient c_k, also a vector of size 64. The transform is fully symmetric. The forward and inverse routines are identical.

```
#define shift 3
void CATransform(int *ck, int *f)
{
        int A,B,C,D,E,F,G,H;
        int AMB,APB,CMD,CPD,EMF,EPF,GMH,GPH;
        A=f[0]+f[1]-(f[2]+f[3]+f[4]+f[5]+f[6]+f[7]);
        B=f[8]+f[9]-(f[10]+f[11]+f[12]+f[13]+f[14]+f[15]);
        C=f[16]+f[17]-(f[18]+f[19]+f[20]+f[21]+f[22]+f[23]);
        D=f[24]+f[25]-(f[26]+f[27]+f[28]+f[29]+f[30]+f[31]);
        E=f[32]+f[33]-(f[34]+f[35]+f[36]+f[37]+f[38]+f[39]);
        F=f[40]+f[41]-(f[42]+f[43]+f[44]+f[45]+f[46]+f[47]);
        G=f[48]+f[49]-(f[50]+f[51]+f[52]+f[53]+f[54]+f[55]);
        H=f[56]+f[57]-(f[58]+f[59]+f[60]+f[61]+f[62]+f[63]);
        AMB=A-B;CMD=C-D;EMF=E-F;GMH=G-H;
        APB=A+B;CPD=C+D;EPF=E+F;GPH=G+H;
        A=APB+CPD;B=EPF+GPH;C=AMB+CMD;D=EMF+GMH;
        E=APB-CPD;F=EPF-GPH;G=AMB-CMD;H=EMF-GMH;
        ck[0]=(-E+B)>>shift;
        ck[8]=(-G+D)>>shift;
        ck[16]=(E+B)>>shift;
        ck[24]= (G+D)>>shift;
        ck[32]= (A-F)>>shift;
        ck[40]= (C-H)>>shift;
        ck[48]= (A+F)>>shift;
        ck[56]= (C+H)>>shift;
        A=f[0]-f[1]-f[2]+f[3]-f[4]+f[5]-f[6]+f[7];
        B=f[8]-f[9]-f[10]+f[11]-f[12]+f[13]-f[14]+f[15];
        C=f[16]-f[17]-f[18]+f[19]-f[20]+f[21]-f[22]+f[23];
        D=f[24]-f[25]-f[26]+f[27]-f[28]+f[29]-f[30]+f[31];
        E=f[32]-f[33]-f[34]+f[35]-f[36]+f[37]-f[38]+f[39];
        F=f[40]-f[41]-f[42]+f[43]-f[44]+f[45]-f[46]+f[47];
        G=f[48]-f[49]-f[50]+f[51]-f[52]+f[53]-f[54]+f[55];
        H=f[56]-f[57]-f[58]+f[59]-f[60]+f[61]-f[62]+f[63];
        AMB=A-B;CMD=C-D;EMF=E-F;GMH=G-H;
        APB=A+B;CPD=C+D;EPF=E+F;GPH=G+H;
        A=APB+CPD;B=EPF+GPH;C=AMB+CMD;D=EMF+GMH;
        E=APB-CPD;F=EPF-GPH;G=AMB-CMD;H=EMF-GMH;
```

Program B.2 *Fast two-dimensional, dual-coefficient CA Transform*

Program B.2 (cont'd)

```
ck[1]=(-E+B)>>shift;
ck[9]=(-G+D)>>shift;
ck[17]= (E+B)>>shift;
ck[25]= (G+D)>>shift;
ck[33]= (A-F)>>shift;
ck[41]= (C-H)>>shift;
ck[49]= (A+F)>>shift;
ck[57]= (C+H)>>shift;
A=f[0]+f[1]-f[2]-f[3]+f[4]+f[5]+f[6]+f[7];
B=f[8]+f[9]-f[10]-f[11]+f[12]+f[13]+f[14]+f[15];
C=f[16]+f[17]-f[18]-f[19]+f[20]+f[21]+f[22]+f[23];
D=f[24]+f[25]-f[26]-f[27]+f[28]+f[29]+f[30]+f[31];
E=f[32]+f[33]-f[34]-f[35]+f[36]+f[37]+f[38]+f[39];
F=f[40]+f[41]-f[42]-f[43]+f[44]+f[45]+f[46]+f[47];
G=f[48]+f[49]-f[50]-f[51]+f[52]+f[53]+f[54]+f[55];
H=f[56]+f[57]-f[58]-f[59]+f[60]+f[61]+f[62]+f[63];
AMB=A-B;CMD=C-D;EMF=E-F;GMH=G-H;
APB=A+B;CPD=C+D;EPF=E+F;GPH=G+H;
A=APB+CPD;B=EPF+GPH;C=AMB+CMD;D=EMF+GMH;
E=APB-CPD;F=EPF-GPH;G=AMB-CMD;H=EMF-GMH;
ck[2]= (E-B)>>shift;
ck[10]= (G-D)>>shift;
ck[18]=(-E-B)>>shift;
ck[26]=(-G-D)>>shift;
ck[34]=(-A+F)>>shift;
ck[42]=(-C+H)>>shift;
ck[50]=(-A-F)>>shift;
ck[58]=(-C-H)>>shift;
A=f[0]-f[1]-f[2]+f[3]+f[4]-f[5]+f[6]-f[7];
B=f[8]-f[9]-f[10]+f[11]+f[12]-f[13]+f[14]-f[15];
C=f[16]-f[17]-f[18]+f[19]+f[20]-f[21]+f[22]-f[23];
D=f[24]-f[25]-f[26]+f[27]+f[28]-f[29]+f[30]-f[31];
E=f[32]-f[33]-f[34]+f[35]+f[36]-f[37]+f[38]-f[39];
F=f[40]-f[41]-f[42]+f[43]+f[44]-f[45]+f[46]-f[47];
G=f[48]-f[49]-f[50]+f[51]+f[52]-f[53]+f[54]-f[55];
H=f[56]-f[57]-f[58]+f[59]+f[60]-f[61]+f[62]-f[63];
AMB=A-B;CMD=C-D;EMF=E-F;GMH=G-H;
APB=A+B;CPD=C+D;EPF=E+F;GPH=G+H;
A=APB+CPD;B=EPF+GPH;C=AMB+CMD;D=EMF+GMH;
E=APB-CPD;F=EPF-GPH;G=AMB-CMD;H=EMF-GMH;
ck[3]= (E-B)>>shift;
ck[11]=(G-D)>>shift;
ck[19]=(-E-B)>>shift;
```

Program B.2 (cont'd)

```
ck[27]=(-G-D)>>shift;
ck[35]=(-A+F)>>shift;
ck[43]=(-C+H)>>shift;
ck[51]=(-A-F)>>shift;
ck[59]=(-C-H)>>shift;
A=f[0]+f[1]+f[2]+f[3]-f[4]-f[5]+f[6]+f[7];
B=f[8]+f[9]+f[10]+f[11]-f[12]-f[13]+f[14]+f[15];
C=f[16]+f[17]+f[18]+f[19]-f[20]-f[21]+f[22]+f[23];
D=f[24]+f[25]+f[26]+f[27]-f[28]-f[29]+f[30]+f[31];
E=f[32]+f[33]+f[34]+f[35]-f[36]-f[37]+f[38]+f[39];
F=f[40]+f[41]+f[42]+f[43]-f[44]-f[45]+f[46]+f[47];
G=f[48]+f[49]+f[50]+f[51]-f[52]-f[53]+f[54]+f[55];
H=f[56]+f[57]+f[58]+f[59]-f[60]-f[61]+f[62]+f[63];
AMB=A-B;CMD=C-D;EMF=E-F;GMH=G-H;
APB=A+B;CPD=C+D;EPF=E+F;GPH=G+H;
A=APB+CPD;B=EPF+GPH;C=AMB+CMD;D=EMF+GMH;
E=APB-CPD;F=EPF-GPH;G=AMB-CMD;H=EMF-GMH;
ck[4]= (E-B)>>shift;
ck[12]=(G-D)>>shift;
ck[20]=(-E-B)>>shift;
ck[28]=(-G-D)>>shift;
ck[36]=(-A+F)>>shift;
ck[44]=(-C+H)>>shift;
ck[52]=(-A-F)>>shift;
ck[60]=(-C-H)>>shift;
A=f[0]-f[1]+f[2]-f[3]-f[4]+f[5]+f[6]-f[7];
B=f[8]-f[9]+f[10]-f[11]-f[12]+f[13]+f[14]-f[15];
C=f[16]-f[17]+f[18]-f[19]-f[20]+f[21]+f[22]-f[23];
D=f[24]-f[25]+f[26]-f[27]-f[28]+f[29]+f[30]-f[31];
E=f[32]-f[33]+f[34]-f[35]-f[36]+f[37]+f[38]-f[39];
F=f[40]-f[41]+f[42]-f[43]-f[44]+f[45]+f[46]-f[47];
G=f[48]-f[49]+f[50]-f[51]-f[52]+f[53]+f[54]-f[55];
H=f[56]-f[57]+f[58]-f[59]-f[60]+f[61]+f[62]-f[63];
AMB=A-B;CMD=C-D;EMF=E-F;GMH=G-H;
APB=A+B;CPD=C+D;EPF=E+F;GPH=G+H;
A=APB+CPD;B=EPF+GPH;C=AMB+CMD;D=EMF+GMH;
E=APB-CPD;F=EPF-GPH;G=AMB-CMD;H=EMF-GMH;
ck[5]= (E-B)>>shift;
ck[13]=(G-D)>>shift;
ck[21]=(-E-B)>>shift;
ck[29]=(-G-D)>>shift;
ck[37]=(-A+F)>>shift;
ck[45]=(-C+H)>>shift;
```

Program B.2 (cont'd)

```
ck[53]=(-A-F)>>shift;
ck[61]=(-C-H)>>shift;
A=f[0]+f[1]+f[2]+f[3]+f[4]+f[5]-f[6]-f[7];
B=f[8]+f[9]+f[10]+f[11]+f[12]+f[13]-f[14]-f[15];
C=f[16]+f[17]+f[18]+f[19]+f[20]+f[21]-f[22]-f[23];
D=f[24]+f[25]+f[26]+f[27]+f[28]+f[29]-f[30]-f[31];
E=f[32]+f[33]+f[34]+f[35]+f[36]+f[37]-f[38]-f[39];
F=f[40]+f[41]+f[42]+f[43]+f[44]+f[45]-f[46]-f[47];
G=f[48]+f[49]+f[50]+f[51]+f[52]+f[53]-f[54]-f[55];
H=f[56]+f[57]+f[58]+f[59]+f[60]+f[61]-f[62]-f[63];
AMB=A-B;CMD=C-D;EMF=E-F;GMH=G-H;
APB=A+B;CPD=C+D;EPF=E+F;GPH=G+H;
A=APB+CPD;B=EPF+GPH;C=AMB+CMD;D=EMF+GMH;
E=APB-CPD;F=EPF-GPH;G=AMB-CMD;H=EMF-GMH;
ck[6]= (E-B)>>shift;
ck[14]=(G-D)>>shift;
ck[22]=(-E-B)>>shift;
ck[30]=(-G-D)>>shift;
ck[38]=(-A+F)>>shift;
ck[46]=(-C+H)>>shift;
ck[54]=(-A-F)>>shift;
ck[62]=(-C-H)>>shift;
A=f[0]-f[1]+f[2]-f[3]+f[4]-f[5]-f[6]+f[7];
B=f[8]-f[9]+f[10]-f[11]+f[12]-f[13]-f[14]+f[15];
C=f[16]-f[17]+f[18]-f[19]+f[20]-f[21]-f[22]+f[23];
D=f[24]-f[25]+f[26]-f[27]+f[28]-f[29]-f[30]+f[31];
E=f[32]-f[33]+f[34]-f[35]+f[36]-f[37]-f[38]+f[39];
F=f[40]-f[41]+f[42]-f[43]+f[44]-f[45]-f[46]+f[47];
G=f[48]-f[49]+f[50]-f[51]+f[52]-f[53]-f[54]+f[55];
H=f[56]-f[57]+f[58]-f[59]+f[60]-f[61]-f[62]+f[63];
AMB=A-B;CMD=C-D;EMF=E-F;GMH=G-H;
APB=A+B;CPD=C+D;EPF=E+F;GPH=G+H;
A=APB+CPD;B=EPF+GPH;C=AMB+CMD;D=EMF+GMH;
E=APB-CPD;F=EPF-GPH;G=AMB-CMD;H=EMF-GMH;
ck[7]= (E-B)>>shift;
ck[15]=(G-D)>>shift;
ck[23]=(-E-B)>>shift;
ck[31]=(-G-D)>>shift;
ck[39]=(-A+F)>>shift;
ck[47]=(-C+H)>>shift;
ck[55]=(-A-F)>>shift;
ck[63]=(-C-H)>>shift;
```

Bibliography

1. Alpert, B., G. Beylkin, R. Coifman, and V. Rodkhlin, Wavelet like bases for the fast solution of second-kind integral equations, SIAM J. Sci. Statist. Comp. 14, pp. 159-184, 1993.
2. Barnsley, M.F., Fractals Everywhere, Second Edition, Academic Press Professional, 1993.
3. Barnsley, M.F. and L.P. Hurd, Fractal Image Compression, AK Peters, Ltd., 1993.
4. Berlekamp, E.R., J.H. Conway and R.K. Guy, *Winning Ways for your Mathematical Plays*, Volume 2, Academic Press, Chapter 25, 1982.
5. Boghosian, B.M., *Cellular Automata Simulation of Two; Phase Flow on the CM-2 Connection Machine Computer*, Technical Report TR-19 CA88-1, Thinking Machines Corporation, 1988.
6. Brieger, L. and E. Bonomi, *A Stochastic Cellular Automaton Simulation of the Nonlinear Diffusion Equation*, Physica D., 47, 1991.
7. Burke, M., *Can CA be used to Model Ecological Systems?*, in H. Gutowitz, editor, Frequently Asked Questions about Cellular Automata; Contributions from the CA Community, Nov. 1994.
8. Cahan, B.D. and O.E. Lafe, *On the Iterative Boundary Element Method*, Computational Engineering with Boundary Elements, 2, Solid and Computational Problems, BETECH90, A.H-D. Cheng,

C.A. Brebbia, and S. Grilli, editors, Computational Mechanics Publishers, Southampton, 1990.

9. Canning, A. and M. Droz, *A Comparison of Spin Exchange and Cellular Automaton Models for Diffusion-Controlled Reactions*, Physica D., 45, 1990.

10. Chen, D., K. Diemer, G.D. Doolen, K. Eggert, C. Fu, S. Gutman and B.J. Travis, *Lattice Gas Automata for Flow through Porous Media*, Physica D., 47, 1991.

11. Cheng, A.H-D. and O.E. Lafe, *Stochastic Boundary Elements for Groundwater Flow with Random Hydraulic Conductivity*, in Boundary Integral Methods, Theory and Applications, L. Morino and R. Piva, editors, Springer-Verlag, Rome, 1990.

12. Chopard, B. and M. Droz, *Cellular Automata Approach to Non-equilibrium Phase Transitions in a Surface Reaction Model: Static and Dynamic Properties*, J. Phys. A 21, 1988.

13. Cliffe, K.A., R.D. Kingdon, P. Schofield and P.J. Stopford, *Lattice Gas Simulation of Free-Boundary Flows*, Physica D., 47, 1991.

14. Cosenza, G., and F. Neri, *Cellular Automata and Nonlinear Dynamical Models*, Physica D., 27, 1987.

15. Creutz, M., *Deterministic Using Dynamics*, Ann. Phys., 167, 1986.

16. Dab, D., J-P., Boon and Y-X., Li, *Lattice-Gas Automata for Coupled Reaction-Diffusion Equations*, Phys. Rev. Lett., 66(19), 1991.

17. Delahaye, J-P., *Les Automates*, Pur La Science, pp.126-134, Nov. 1991.

18. Despain, A., C.E. Max, G. Doolen and B. Hasslacher, *Prospects for a Lattice-Gas Computer*, in Lattice Gas Methods for Partial Differential Equations, Santa Fe Institute Studies in the Sciences of Complexity, Eds. Doolen *et al.*, Addison-Wesley, pp.211-218, 1990.
19. d'Humiéres, D. and P. Lallemand, *Numerical Solutions of Hydrodynamics with Lattice Gas Automata in Two Dimensions*, Complex Systems, 1, pp.599-632, 1987.
20. Di Pietro, L.B., A. Melayah and S. Zaleski, *Modeling Water Infiltration in Unsaturated Porous Media by Interacting Lattice Gas-Cellular Automata*, Water Resource Res., 30(10), 1994.
21. Ernst, M.H. and Das, S.P., *Thermal Cellular Automata Fluids*, Journal of Statistical Physics, 66(1/2), 1992.
22. Fisher, Y. (ed.), Fractal Image Compression: Theory and Applications, Springer-Verlag, 1995.
23. Frisch, U., B. Haaslacher and Y. Pomeau, *Lattice-Gas Automata for the Navier-Stokes Equation*, Phys. Rev. Lett., 56(14), 1986.
24. Frisch, U., D. d'Humiéres, B. Hasslacher, P. Lallemand, Y. Pomeau and J-P. Rivet, *Lattice Gas Hydrodynamics in Two and Three Dimensions*, Complex Systems 1, pp.649-707, 1987.
25. Frisch, U., *Relation between the Lattice Boltzmann Equation and the Navier-Stokes Equations*, Physica D, 47, pp.231-232, 1991.
26. Garcia-Ybarra, P.L., A. Lopez-Martin and J.C. Antoranz, *Unsteady Potential Flows Computation by Cellular Automata: The Premixed Flame Instability*, Transport Theory and Statistical Physics, 23(1/3):173, 1994.

27. Gonzalez, R.C. and R.E. Woods, *Digital Image Processing*, Addison-Wesley Publishing Company, New York, 1992.

28. Grassberger, P., *A New Mechanism for Deterministic Diffusion*, Wuppertal preprint WU B, 82-18, 1982.

29. Green, D.G., *Simulated Effects of Fire, Dispersal and Spatial Pattern on Competition within Forest Mosaics*, Vegetation, 82, pp.139-154, 1982.

30. Green, D.G., A.P.N. House and S.M. House, *Simulating Spatial Patterns in Forest Ecosystems*, Mathematics and Computers in Simulation, 27, pp.191-198, 1985.

31. Greenberg, J.M. and S.P. Hastings, *Spatial Patterns for Discrete Models of Diffusion in Excitable Media*, SIAM J. Appl. Math., 34, 1978.

32. Greenberg, J. M., B.D. Hassard and S.P. Hastings, *Pattern Formation and Periodic Structures in Systems Modeled by Reaction-Diffusion Equations*, Bull. Am. Math. Soc., 84, 1978.

33. Greenberg, J.M., C. Greene and S. Hastings, *A Combinatorial Problem Arising in the Study of Reaction-Diffusion Equations*, SIAM J. Alg. Disc. Meth., 1, 1980.

34. Guan, P., *Cellular Automation Public-Key Crytosystems*, Complex Systems, 1, 1987.

35. Gunstensen, A.K. and D.H. Rothman, *A Lattice Gas Model for Three Immiscible Fluids*, Physica D., 47, pp.47-52, 1991a.

36. Gunstensen, A.K. and D.H. Rothman, *A Galilean-Invariant Immiscible Lattice Gas*, Physica D., 47, pp.53-63, 1991b.

37. Gutowitz, H.A., *A Hierarchical Classification of Cellular Automata*, Physica D., 45, 1990.

38. Gutowitz, H., *Method and Apparatus for Encryption, Description and Authentication using Dynamical Systems,* U.S. Patent #5,365,589, Issued Nov. 15, 1994.

39. Gutowitz, H., *Cryptography with Dynamical Systems,* in N. Boccara, E. Goles, S. Martinez and P. Picco, editors, Cellular Automata and Cooperative Phenomena, pp.237-274, Kluwer Academic Publishers, 1993a.

40. Hartman, H. and P. Tamayo, *Reversible Cellular Automata and Chemical Turbulence,* Physica D., 45, 1990.

41. Hillis, W.D., *The Connection Machine: A Computer Architecture based on Cellular Automata,* Physica D., 45, 1990.

42. Hogeweg, P., *Cellular Automata as a Paradigm for Ecological Modeling,* Applied Mathematics and Computation, 27, pp.81-100, 1988.

43. Howard, N., R. Taylor and N. Allinson, *The Design and Implementation of a Massively Parallel Fuzzy Architecture,* Proc. IEEE, pp.545-552, 1992.

44. Huberman, B.A. and N. Glance, *Evolutionary Games and Computer Simulations,* Proc. Natl. Acad. Sci. USA, 90, pp.7716-7718, 1993.

45. Kohring, G.A., *Calculations of Drag Coefficients via Hydrodynamic Cellular Automata,* Jour de Physique, 2(3), 1992a.

46. Kohring, G.A., *The Cellular Automata Approach to Simulating Fluid Flows in Porous Media,* Physica A, 186, 1992b.

47. Kohring, G.A., *An Efficient Hydrodynamic Cellular Automata for Simulating Fluids with Large Viscosities,* Journal of Statistical Physics, 66(3/4), 1992c.

48. Lafe, O.E., *Data Compression and Encryption Using Cellular Automata Transforms*, Engineering Applications of Artificial Intelligence, 10, pp. 581-591, 1997.

49. Lafe, O.E., *Method and Apparatus for Data Encryption/Decryption Using Cellular Automata Transform*, U.S. Patent #5,677,956, Issued Oct. 14, 1997.

50. Lafe, O.E. and A.H-D. Cheng, *Stochastic Analysis of Groundwater Flow*, in D. Benasari, C.A. Brebbia, and D. Ouazar, editors, in Computational Water Resources, Computational Mechanics Publications, Southampton Boston, 1991.

51. Lafe, O.E. and A.H-D. Cheng, *Stochastic Indirect Boundary Element Method*, in Computational Stochastic Mechanics, A.H-D Cheng and C.Y. Yang, editors, Computational Mechanics Publications, Southampton Boston, 1993.

52. Langton, C.G., *Studying Artificial Life with Cellular Automata*, Physica D., 22, pp.120-149, 1986.

53. Margolus, N. and T. Toffoli, *Cellular Automata Machines*, in Lattice Gas Methods for Partial Differential Equations, Santa Fe Institute Studies in the Sciences of Complexity, Eds. Doolen *et al.*, Addison-Wesley, pp.219-249, 1990.

54. Orszag, S.A. and V. Yakhot, *Reynolds Number Scaling of Cellular-Automaton Hydrodynamics*, in Lattice Gas Methods for Partial Differential Equations, Santa Fe Institute Studies in the Sciences of Complexity, Eds. Doolen *et al.*, Addison-Wesley, pp.269-273, 1990.

55. Perera, A., K.A. Penson and U. Schultze, *Long-time Dynamics of Two-dimensional Fluid Binary Mixture in Cellular Automata Models,* Helvetica Physica Acta, 65(2/3), 1992.

56. Pohlmann, K.C., Principles of Digital Audio, McGraw-Hill, Inc., 1995.

57. Poundstone, W., The Recursive Universe, William Marrow and Company, NY, 1985.

58. Richards, F.C., T.P. Meyer and N.H. Packard, *Extracting Cellular Automaton Rules Directly from Experimental Data,* Physica D., 45, pp.189-202, 1990.

59. Rothman, D.H., *Cellular-Automaton Fluids: A Model for Flow in Porous Media,* Geophysics, 53, pp.509-518, 1988.

60. Rothman, D.H. and J.M. Keller, *Immiscible Cellular-Automaton Fluids,* J. Stat. Phys., 52, pp.1119-1127, 1988.

61. Schneier, B., Applied Cryptography, John Wiley & Sons, New York, 1993.

62. Silvertown, J., S. Holtier, J. Johnson and P. Dale, *Cellular Automaton Models of Interspecific Competition for Space - the Effect of Pattern on Process,* Journal of Ecology, 80, pp.527-534, 1992.

63. Toffoli, T., *Cellular Automata as an Alternative to (Rather than an Approximation of) Differential Equations in Modeling Physics,* Physica D., 10, pp. 117-127, 1984a.

64. Toffoli, T., *CAM: A High-Performance Cellular-Automaton Machine,* Physica D., 10, pp. 195-204, 1984b.

65. Toffoli, T., and N. Margolus, Cellular Automata Machines - A new Environment for Modeling, The MIT Press, Cambridge (Mass), pp. 8-11, 1987.

66. Toffoli, T. and N.H. Margolus, *Invertible Cellular Automata*, Physica D., 45, pp. 229-253, 1990.

67. Toffoli, T. and N.H. Margolus, *Programmable Matter: Concepts and Realization*, Physica D., 47, pp. 263-272, 1991.

68. Toffoli, T., *Occam, Turing, von Neumann, Jaynes: How much caan you get for how little? (A Conceptual Introduction to Cellular Automata)*, InterJournal, Dec., 1994.

69. Toffoli, T., *Cellular Automata*, in Handbook of Brain Theory and Neural Networks, (Michael Arbib ed.), MIT Press, pp. 166-169, 1995.

70. Toffoli, T., *Programmable Matter Methods*, in Future Generation Computer Systems, special issue on "Cellular Automata: Promise and Prospects in Computational Science" (D. Talia and P. Sloot ed.), 1998.

71. Ulam, S.M., *On Some Mathematical Problems Connected with Patterns of Growth of Figures*, Proceedings of Symposia in Applied Mathematics, 14, pp.215-224, 1962.

72. Vichniac, G.Y., *Simulating Physics with Cellular Automata*, Physica D., 10, 1984.

73. Wells, J.T., D.R. Janecky and B.J. Travis, *A Lattice Gas Automata Model for Heterogeneous Chemical Reactions at Mineral Surfaces and in Pore Networks*, Physica D., 47, 1991.

74. Weimer, J.R. and J-P. Boon, *Class of Cellular Automata for Reaction-Diffusion Systems*, Physical Review E, 49(2), 1994.

75. Wolfram, S., *Statistical Mechanics of Cellular Automata*, Rev. Mod. Phys., 55(3), 1983.

76. Wolfram, S., *Cryptography with Cellular Automata*, Proceedings of Crypto '85, pp.429-432, 1985.

77. Yakhot, V., J. Bayly and S.A. Orszag, *Analogy between Hyperscale Transport and Cellular Automaton Fluid Dynamics*, in Lattice Gas Methods for Partial Differential Equations, Santa Fe Institute Studies in the Sciences of Complexity, Eds. Doolen *et al.*, Addison-Wesley, pp.283-288, 1990.

Index